Choice, Not Fate

Choice, Not Fate

Shaping a Sustainable Future in the Space Age

James A. Vedda, Ph.D.

Copyright © 2009 by James A. Vedda.

Library of Congress Control Number: 2009913427
ISBN: Hardcover   978-1-4500-1348-2
               Softcover   978-1-4500-1347-5
               Ebook   978-1-4500-1349-9

All rights reserved. No part of this book may be reproduced or transmitted in any form or by any means, electronic or mechanical, including photocopying, recording, or by any information storage and retrieval system, without permission in writing from the copyright owner.

This book was printed in the United States of America.

Cover photo: NASA

To order additional copies of this book, contact:
Xlibris Corporation
1-888-795-4274
www.Xlibris.com
Orders@Xlibris.com

Contents

Acknowledgements ix

Introduction xi

Chapter 1: Cruising to Utopia—or Not 1

Chapter 2: Searching for a Vision of the Future 20

Chapter 3: Muddling Through with a
Short-Term View 46

Chapter 4: The Bureaucracy: Best Hope for
the Future? 78

Chapter 5: Astropreneurs: The Real Vision,
or Just a Dream With Good Special
Effects? 98

Chapter 6: Be Careful What You Wish For 126

Chapter 7: Earth as an Open System 146

Chapter 8: The Century Perspective 168

Chapter 9: Commitment to the Future 183

Bibliography 193

Dedication

To my mentor and lifelong friend, David C. Webb

Acknowledgements

I would like to thank my colleagues who reviewed the manuscript and provided comments: Dr. David C. Webb, space consultant and educator; Dr. Joan Johnson-Freese, chair of the National Security Decision Making Department at the Naval War College; and Dr. Roger Launius, curator of space history at the Smithsonian National Air & Space Museum.

I'm also thankful to my wife Lin, who tolerated me for several months of evenings and weekends during which I was distracted by research and writing.

Introduction

The dogmas of the quiet past are inadequate to the stormy present. The occasion is piled high with difficulty, and we must rise with the occasion. As our case is new, so we must think anew and act anew. —Abraham Lincoln

President Lincoln was speaking in 1862 of a divided nation, at war with itself over issues of great and lasting consequence. Although the United States is not threatened by civil war today, it is facing an assortment of issues of great and lasting consequence for the nation and the world. In that context, Lincoln's words convey a message that remains valid today as we contemplate which near-term actions will yield some desired future.

I titled this book *Choice, Not Fate* to emphasize that the future is what we make it. Persistent trends are important indicators of what the future holds, and there's always the possibility of sudden, disruptive change, but we're not powerless to exert influence over our individual and collective futures, especially if we act early and think creatively.

In order to have a positive influence on our future, we obviously need to set goals, identify problems, formulate workable solutions, and implement them so we reach the desired endpoint in a timely manner while minimizing unpleasant consequences. Conceptually, that sounds simple, logical, and so obvious as to be hardly worth mentioning. On very large scales, however, the problems become highly complex, resulting in a multitude of obstacles and a lack of adequate tools and resources. The impediment we most often hear about is that we don't have enough money to do everything we want, but there's more to it than that, and this book will explore some of these other difficulties. My aim is to shed light on what I believe is a core problem: the cultural and institutional forces that drive us to short-term thinking at a time when we desperately need more and better long-term strategic planning.

Over the past four decades, there have been plenty of books and articles presenting gloomy scenarios for the environment, the economy, traditional cultures, and even the whole of human society. I've read quite a few of them, and yet I remain an optimist. Okay, a cautious optimist. Although I quote from some of these writings in the chapters ahead, it's not my intention to present a lengthy restatement of the same message.

Rather, I want to highlight the roadblocks to sustainable, positive action, thereby encouraging exploration of different ways of thinking about the future, because the path to a favorable future, I believe, will be a challenging and unconventional one.

The focal point of this discussion is the exploration and development of space—the quintessential example of the long-term perspective. I believe space will be a major factor in enabling a favorable future, but only if it becomes a mainstream activity that contributes substantially to addressing global problems, growing the world economy, and expanding human horizons in science, technology, and society. In addressing the possibilities and pitfalls of long-term strategic planning for space, this book does not provide an overview of the latest NASA programs or speculate on the next generation of space launch vehicles, tasks that have been well served by other authors. Instead, these chapters assess our attempts to build a path to the future, and point out some of the many potholes on the road we travel. In this analysis, it's necessary to touch on issues like globalization, climate change, and the interactions of government entities, which at first may seem unrelated to the subject of space. But in reality they are closely related, and it's important for those who are involved or interested in space development to resist the temptation to ignore them. Space activities often are misunderstood, viewed as a luxury undertaken by wealthy countries in good economic times, somehow disconnected from the rest of the world's concerns. On the contrary, even though most people don't realize it yet, space is intimately linked to a vast array of human activities. The last three chapters suggest a way of thinking about the exploration and development of space that takes this into account and differs from the conventional approach of recent decades.

The points made in this book regarding the need for more and better long-term thinking are applicable to other activities besides space development. Cultural and institutional barriers like the ones confronting space efforts are present in both the public and private sectors across a wide range of activities on which our future hinges. My tendency, when I read thought-provoking writings about prospects for our future, is to ask, "How is this interrelated with space development?" Similarly, I hope that as you read this book, you'll be contemplating how the issues presented here relate to your professional and personal interests.

Regarding references to various timeframes, I prefer to think of the "short term" as the next 10 to 20 years, the "medium term" as the middle decades of the 21st century, and the "long term" as the late 21st

century and beyond. Clearly, not everyone uses the terms in this way. For example, some view "long term" as anything beyond the next election or the next budget cycle. But even if you don't expect to be around for (my definition of) the long term, I'm assuming that you have friends, relatives, or descendents who will be. You also may have given some thought to non-trivial topics like the survival of our species and improvement of the human condition. Given the challenges ahead, these are more than just academic subjects, and therefore are worthy of serious and immediate consideration.

Chapter 1
Cruising to Utopia—or Not

> *It is change, continuing change, inevitable change, that is the dominant factor in society today. No sensible decision can be made any longer without taking into account not only the world as it is, but the world as it will be . . . This, in turn, means that our statesmen, our businessmen, our everyman must take on a science fictional way of thinking.*—Isaac Asimov

As a baby boomer growing up in America, I came of age believing that things would steadily get better, at least for the rest of my lifetime. There would be ups and downs, of course, but the overall trend for me, for my country, and for the world, would be positive. Technology would advance, the standard of living would improve for most people, security would increase at home and abroad. To a great extent, this world view seemed to be unfolding quite nicely through the end of the 20th century. Impressive advances were occurring on a regular basis in information technologies, medical science, and many other fields. By the 1990s, the Cold War threat of superpower confrontation had ended; the U.S. economy was healthy, with low unemployment, low inflation, and steady growth; the U.S. crime rate was down; and by the end of the decade, the U.S. federal budget had a surplus for the first time in 30 years. Globalization had become the new international model, and it was going to bring the world together and help it prosper.

But as we entered the 21st century, it became apparent that some negative developments were more than just bumps in the road: slackening of U.S. investment in important technologies; deficiencies in the nation's education system in general and its technical workforce in particular; escalating conflict in many parts of the world, much of it perpetrated by non-state actors; increasing disparity between haves and have-nots; a global economic downturn; and the looming threat of climate change, with its timing and severity unknown. These conditions did not appear suddenly. Rather, they were rooted in numerous unsustainable industrial, financial, and consumption practices that had contributed to the successes of previous decades while shortchanging the future.

Clearly, we're not simply fated to succeed. The U.S. and its allies defeated Nazi fascism and contained Soviet expansionism in the 20th century and in the process became economically, technologically, and militarily powerful. But that doesn't mean we can put our society on cruise control until we reach utopia. The world still needs plenty of tough love, and the forward momentum of our recent past needs to be recharged if we're going to be up to the task. That means making farsighted choices and sticking to them.

News from the future: a call to action (or depression)
The list of the world's problems seems endless judging from our daily exposure to various news media, offerings at the local bookstore, gossip around the water cooler, and the pontifications of barbers and cab drivers. The difficulty, in an environment of limited political consensus and even more limited resources, is deciding which problems are addressable, which ones get first priority, which solutions are best, and how they are going to be paid for. Fortunately, numerous global trend-watchers have published their observations and prognostications in attempts to address some or all of these concerns. A very small sample of them are highlighted here, enough to pick out the recurring themes.

In 1972, as the Apollo lunar missions were ending and environmental awareness was growing, the international think tank The Club of Rome released a study, conducted by a group based at the Massachusetts Institute of Technology (MIT), called *The Limits to Growth*. The report received wide notoriety, and much derision from some quarters, for what was seen as a doomsday prediction:

> If the present growth trends in world population, industrialization, pollution, food production, and resource depletion continue unchanged, the limits to growth on this planet will be reached sometime within the next one hundred years. The most probable result will be a rather sudden and uncontrollable decline in both population and industrial capacity.

As often happens when findings such as this get reported in the media, the report's other, less-sensational conclusion was largely ignored: that it is possible to alter these trends.

Within a few years after the study's release, its many detractors appeared to have convinced general audiences that its conclusions had been debunked and its methodology discredited. I found it puzzling that critics were so quick to dismiss this work, declaring the 100-year scenarios disproven because no global societal or industrial collapse had occurred in the decade after 1972. According to my calendar, we're still a long way from 2072. In the meantime, the things we've learned about population trends, resource supply and demand, and threats to the environment and climate have shown us that rather than being disproven, the study's scenarios are looking ominously accurate.

While I was puzzled at the dismissive and sometimes belligerent responses to the report, its authors must have been deeply frustrated. That may have been what prompted them to do updates roughly every 10 years incorporating new data and more sophisticated mathematical modeling. (Their 30-year update was released in 2004.) Perhaps the most critical shortcoming of *The Limits to Growth* and the series of updates is one the authors openly acknowledge. They address flows of population, materials, energy, and emissions that can be mathematically modeled, but do not include factors such as military conflict, rampant political corruption, natural disasters, pandemics, or severe economic stresses like currency and debt crises. If these things are taken into account, one could view the *Limits to Growth* model as wildly optimistic.

For purposes of this discussion, the most important thing to take away from *The Limits to Growth* is the list of what were judged to be the five major trends of global concern: accelerating industrialization, rapid population growth, widespread malnutrition, depletion of nonrenewable resources, and a deteriorating environment. Though daunting, these concerns are being addressed to a limited extent today, but we need to accelerate our capacity to address them in the future.

As someone who has been observing and studying the exploration and development of space since the age of six, and has spent the past quarter-century trying to contribute to it professionally, I naturally think about how space capabilities can contribute, and in some cases already are contributing, to global solutions. This compels me to highlight a key deficiency that can be found in *The Limits to Growth* and every other study of global problems and proposed solutions: the assumption that the Earth is a closed system—nothing (other than light) goes in or out.

This assumption should be kept in mind as we review what others have written about the planet's problems. By the end of the book, I hope to demonstrate that we should change this assumption and start treating the Earth as an open system. Of course, a lot of things in our culture, our institutions, our goals, and our priorities will need to evolve if we are to achieve this. Following chapters will address these issues; for now, let's examine more insights from those who have been diagnosing the world's ills.

In 1993, Yale historian Paul Kennedy published *Preparing for the Twenty-First Century* in which he examined global problems and trends with an eye toward the interactions of circumstances that make societies collapse. As author of the best-selling *The Rise and Fall of the Great Powers*, Kennedy was well versed in how this had occurred historically. His list of troublesome and interrelated trends included:

- Population explosion and changing demographics
- Globalization of business driven by technological advances
- Widespread environmental damage
- Global warming

Note how closely these align with the *Limits to Growth* list. Although Kennedy did not specifically mention malnutrition or resource depletion in his list, he clearly saw them as key components of the problem. Writing at a time when popular recognition of globalization was just beginning to emerge, Kennedy was particularly concerned that the interaction between growing population, job displacement, and illegal migration would widen the rift between rich and poor countries.

More recently, Thomas Homer-Dixon, a professor of political science at the University of Toronto, published *The Upside of Down: Catastrophe, Creativity, and the Renewal of Civilization*, which reads like a post-9/11 update to Kennedy's book. Homer-Dixon refers to the big global problems as tectonic stresses, and lists them as follows:

- Population stress arising from differences in the population growth rates between rich and poor societies, and from the spiraling growth of megacities in poor countries.
- Energy stress—above all from the increasing scarcity of conventional oil.
- Environmental stress from worsening damage to our land, water, forests, and fisheries.
- Climate stress from changes in the makeup of our atmosphere.

- Economic stress resulting from instabilities in the global economic system and ever-widening income gaps between rich and poor people.

Homer-Dixon gives us unmistakable evidence that he sees the Earth as a closed system. He specifically points out that humanity "can't get its resources or expel its pollution beyond Earth's boundaries."

Vaclav Smil, a professor at the University of Manitoba whose interests span several disciplines, produced a book in 2008 with the cheery title *Global Catastrophes and Trends: The Next Fifty Years*. It's a very educational assessment of all the things that could get nasty on a planet-wide scale in less than one lifetime. Smil's message, if I may summarize it briefly, is that we have a long list of things to worry about, but the complexity of the circumstances is such that there's no point in trying to predict what will happen or how quickly it will happen. His list of threats resembles those of many other commentators. He surveys a substantial number of studies on environmental stress, climate change, energy, biodiversity, population growth, demographic changes, and infectious diseases, and even considers geologic events and asteroid impacts. Smil is not willing tell us how rosy or grim he thinks the future will be, citing a lack of evidence that is both quantifiable and consistent, but suffice it to say that he's provided a multidisciplinary feast for those who delight in worst-case scenarios.

This small sample of the literature of recent decades is enough to show a consistent message: the planet suffers from environmental degradation, climate instability, current or anticipated scarcities of energy and raw materials (including food and water), and increasingly uneven distribution of wealth. These problems stimulate conflict and are rooted in the size and density of the world's population and in poorly managed industrial growth.

Globalization: the problem or the solution?

Globalization has been called the dominant trend that has replaced the Cold War. "Replaced" isn't really the right term because it gives the impression that globalization didn't exist prior to the 1991 breakup of the Soviet Union. Actually, the Cold War and the current era of globalization ran concurrently starting with the global economic recovery that followed World War II. Before we explore how this "dominant trend" relates to things like futurism and space development, some discussion is needed on the characteristics of globalization to help us assess its implications, both good and bad, for future actions.

Hundreds of books and countless articles on the subject of globalization have been written since the 1990s. Although there is no consistent definition for the term, you hear the word "globalization" associated with everything from politics to economics to culture to the environment. The economic and societal developments that today are labeled globalization have been around for centuries, waxing and waning at least since the 16th century. Some pundits believe the term's recent popularity is partly due to its ambiguity—it can assume different connotations depending on who is using it and in which context. For some, it connotes international connectedness, liberation from geographic and nationalistic limits to innovation and growth, leveling of inequalities, improvement of living standards and the human condition, and an avenue for avoiding major conflicts like the wars of the 20th century. For others, such as Vandana Shiva of the International Forum on Globalization, it is just the opposite:

> ... a project for polarizing and dividing people—along axis of class and economic inequality, axis of religion and culture, axis of gender, axis of geographies and regions ... a new caste system ...

The deep and often heated disagreement over the nature and ramifications of globalization makes it difficult to find a generally accepted definition for the phenomenon that *New York Times* writer Thomas Friedman has called "the overarching international system shaping the domestic politics and foreign relations of virtually every country." For our purposes, we need a definition that recognizes the contributions of technological development. Here's a good one from Joseph Stiglitz, a Nobel laureate in economics, from his 2003 book *Globalization and Its Discontents*:

> ... the closer integration of the countries and peoples of the world which has been brought about by the enormous reduction of costs of transportation and communication, and the breaking down of artificial barriers to the flows of goods, services, capital, knowledge, and (to a lesser extent) people across borders.

Among the most recognized books that have popularized the concept of globalization are two best-sellers by Thomas Friedman, *The Lexus and the Olive Tree* (1999) and *The World Is Flat* (2005). In the latter, Friedman divides globalization into three distinct eras:

- Globalization 1.0 (1492-1800). Countries and governments drove global integration. Trade began between the Old World and New World.
- Globalization 2.0 (1800-2000, interrupted by the Great Depression and the two World Wars). The Industrial Revolution and multinational companies were the key agents of change.
- Globalization 3.0 (2000 onward). Individuals have newfound power to collaborate and compete globally.

Friedman's view of globalization history hinges on the increasing empowerment of ever-smaller components of societies. This view, though compelling, is not shared by historians, who perceive the three eras of globalization as follows:

- The age of exploration and colonization from the 15th century to the early 19th century.
- Industrialization and expansion of world trade from the mid-19th century to 1914, at which time globalization was halted by the outbreak of the first World War.
- The current era from the post-World War II recovery of the global economy to the present.

When one looks at the differences between the pre-war and post-war eras, it is clear that this latter view is more accurate—the World Wars and the Great Depression separated two distinct eras rather than simply being a pause within a single continuous period. This view is also better suited to the analysis of the role of space development in the current era.

Before looking at the differences between the current and previous experiences with globalization, let's look at characteristics that are similar. For example, 19th century globalization featured the following: unprecedented international movement of capital, raw materials, and people; revolutionary technological innovation, including the telephone, radio, and internal combustion engine; and an ongoing struggle for balance between protectionism and free trade. Allowing for a century of technological advances, these characteristics sound very familiar today. There were also tensions in that era's international order that resemble modern headlines: imperial overstretch, great power rivalry, an unstable alliance system, rogue regimes sponsoring terror, and the rise of a revolutionary terrorist organization (the Bolsheviks in Russia) hostile to capitalism.

The end of World War II launched the current era by allowing the economies of the industrialized nations to recover and grow, particularly in the United States, which had abundant resources and didn't have its homeland demolished in the war. An important legacy of the war that would stimulate the re-emergence of globalization was the new relationship between the U.S. government and the research community. Always a key supporter of infrastructure projects, the U.S. government became the nation's primary patron of science and engineering. An effort to sustain this relationship beyond the war years was spearheaded by the scientific and technical communities that obviously wanted to continue receiving government support, and by political notables such as President Franklin Roosevelt's director of scientific research and development, Vannevar Bush, who wrote a landmark report to the president in 1945 titled "Science—The Endless Frontier." The result was what some have called a "social contract with science" that portrays the pursuit of scientific knowledge as intrinsically good and useful. According to this philosophy, as long as the nation maintains its input into the reservoir of knowledge, the system is working as it should, and application of that knowledge will take care of itself. Institutions created in this image, such as the National Science Foundation and NASA, persist to this day, as does the dominance of government funding in certain fields, such as medical research. However, this social contract has become more difficult to sustain in an evolving post-Cold War political and economic environment. Large science budgets, including those for space, are increasingly difficult to justify if their objectives are perceived to be isolated from societal needs.

The output of this government partnership with science was the eventual widespread availability of technologies that could only be dreamed of, or in some cases were unimaginable, in the prior era of globalization. The results were technologies that we take for granted today which enabled more rapid movement of people, goods, and information—essential to globalization—in the latter half of the 20th century:

- Jet air transport for passengers and cargo multiplied the speed of long-distance travel, effectively shrinking travel times from days to hours. Equally important, it eventually became affordable to a broad swath of society.
- Near-instantaneous, high-bandwidth communications have evolved so far beyond the telegraph and radio of our great-grandfathers' day that the benefits are beyond our ability to quantify. By the 1960s, telephones became ubiquitous and continents were connected by undersea cables. The fax machine became popular in the mid-1980s, and at the same time so-called microcomputers were maturing. By the 1990s, the expectation was that every desktop would have its own computer, probably linked to a corporate network and the Internet.
- Space technology in its various forms started making its contribution to globalization in the 1960s.

International economics scholar and former Secretary of Labor Robert Reich noted in his book *The Future of Success* that although technology and globalization are often discussed as separate trends, they are becoming one and the same. The literature on globalization, however, tends to give space technology cursory treatment, if it's mentioned at all. Typically, all that's said are a few words acknowledging the role of satellite communications as a component of the telecommunications revolution that enabled the current era of globalization. Such limited discussion is understandable, since most of the literature focuses on international economics, implications for developing countries and the world's poor, and potential impacts on the environment. But there is much more to the story of how space development has made the current globalization experience different from previous ones, and will continue to affect its evolution.

During much of the post-war era, U.S. government space programs had fairly well-defined roles that were closely associated with the nation's Cold War interests: national security tools, sources of national prestige, drivers of technological advancement. If globalization is the successor to the Cold War paradigm, then U.S. space efforts, particularly those involving exploration and development, must be redefined in terms of their new globalization-era identity. This is not a simple task because, as noted earlier, debates still rage as to what globalization means, where it is headed, and whether the net effect will be good or bad.

Although globalization debates primarily address economic, social, and environmental issues, the continuing influence of space development cannot be ignored or viewed in isolation from these issues.

Clearly, the contribution of space technology has not been limited to the addition of satellites to an already expanding network of global communications, as the globalization literature seems to imply. The full array of emerging space capabilities had significant influence. For example, numerous business and government activities at the local, regional, national, and international level are dependent on the weather. The improved weather forecasts enabled by satellites beginning in the 1960s improved productivity and safety of operations in areas such as agriculture, air transport, shipping, construction, mining, and utilities, to name a few. Over time, these improvements had cumulative effects that altered business cycles and planning to reflect an evolving information age economy.

As weather monitoring matured, another form of Earth monitoring known as satellite remote sensing became available to civilian users starting in the 1970s. Able to produce images much more detailed than weather satellites, often using multiple bands of the electromagnetic spectrum that reveal even more information, remote sensing opened new avenues for industries like those listed above and for other users, such as urban planners, environmentalists, fossil fuel geologists, and even archeologists. Early in NASA's Landsat series of spacecraft, interest in this new capability spread around the world. Assisted by the U.S. policy of non-discriminatory access to Landsat data—all imagery was available to all interested parties around the world on the same terms—remote sensing became a new tool for resource conservation and exploitation, environmental stewardship, and disaster assistance, among other applications. Commercial descendants of Landsat are cultivating global markets in ventures that are very much in keeping with the proliferation of know-how and exchange of data that are characteristic of a globalized world.

The use of satellites for navigation began in the 1960s, primarily to serve the needs of military ships and submarines. Today, GPS has become a household word (or more appropriately, a household acronym) even to those who don't know that it stands for Global Positioning System. By the end of the 1980s, the GPS constellation was taking shape, and its services—accurate positioning, navigation, and timing—were being shared at no cost with the world. Those services provide value to anything that moves, and even some things that don't move but depend on precise timing signals. Though it's often overlooked, this capability is in a class with satellite communications as an enabler of globalization. Its ability to assist the movement of people, goods, capital, and information around the world is widely recognized, as evidenced by several global and regional navigation satellite systems operated or planned by Europe, Russia, China, Japan, and India.

One of the most important but least acknowledged contributions of satellites to globalization is their role in keeping the Cold War from turning into a hot war. As noted earlier, the previous era of globalization ended abruptly with the outbreak of major military conflict in 1914. Despite the best efforts of many nations, it took more than three decades to resurrect globalization. The same long delay could have occurred in the years following World War II, dramatically worsened by the introduction of nuclear weapons. But the recovery was relatively quick and hugely successful, helped by post-war investments such as the Marshall Plan to reinvigorate Europe's economy. This could only have happened in a relatively stable and peaceful environment that allowed international trade to thrive. The bulk of the credit for preventing World War III from disrupting this environment is generally given to the "balance of terror" in offensive weapons between the United States and the Soviet Union. But the nuclear arsenal could not have allowed us to achieve this without the support of satellites for surveillance, reconnaissance, and targeting. The hair trigger of nuclear weapons was never pulled because each side could use satellites to see what the other had deployed, observe their behavior, and provide early warning of attack.

Present-day globalization is reaping the benefits of space applications created and disseminated in the Cold War in an environment that kept major threats at bay and allowed global markets to flourish. Government space efforts aimed at national security, national prestige, and technology development have led us to a point where civil and commercial space applications are fundamental—though often transparent—in a globalizing world.

Space development under a globalization paradigm

In *The Lexus and the Olive Tree*, Thomas Friedman stated his belief that the current system of globalization "has come upon us far faster than our ability to retrain ourselves to see and comprehend it." Certainly this has been the case with space development. As in other societal activities, space-related institutions seek to continue their existence and their traditional priorities despite the fast pace of change in key segments of their environment.

A significant portion of the established space industry is designed to serve governments, since these constitute much of the customer base in key areas such as space hardware manufacturing and launch services. The relatively small number of competitors and customers in these areas, and the dominance of government customers, yield a space industry that is slower to adapt and innovate than most other high-tech industries. The tendency to protect space technologies as sensitive national assets slows their adoption in the world market and may hinder the competitiveness of nations employing excessive export restrictions and protectionist measures. This does not bode well for the U.S. space community's ability to rapidly adapt to the globalized environment.

In addition to keeping up with the frenetic pace of the world's economic and technical evolution, the U.S. space community also must adapt to other types of changes. In the globalization era, this includes taking into account the diffusion (or "democratization") of technology, information, economic power, and international influence. The leveling effect that results will change relationships with international partners, increase competition in space products and services on the world market, and challenge U.S. space leadership across the board. This is already forcing the U.S. civil space program to rethink its post-Cold War identity, as demonstrated by NASA's shift of focus away from the space shuttle and space station, its flagship programs of the 1970s and 80s.

As before, geopolitics and economics will drive the search for a new national identity in space. The issues are somewhat different today than they were during the Cold War, but the challenges and risks remain. The International Forum on Globalization warns of this when it states that: "The world's corporate and political leadership is undertaking a restructuring of global politics and economics that may prove as historically significant as any event since the Industrial Revolution. This restructuring is happening at tremendous speed, with little public disclosure of the profound consequences affecting democracy, human welfare, local economies, and the natural world." Warnings of this type are echoed by many observers of globalization, some of whom see the movement as unsustainable, sparking non-linear trends that will be impossible to manage. In a 2008 assessment titled *Global Trends 2025*, the U.S. intelligence community's National Intelligence Council showed less pessimism than the globalization critics, but still took a cautionary tone:

> History tells us that rapid change brings many dangers... we do not believe that we are headed toward a complete breakdown—as occurred in 1914-1918 when an earlier phase of globalization came to a halt. However, the next 20 years of transition toward a new international system are fraught with risks—more than we envisaged [in the previous trend study in this series, published in 2004].

Supporters of globalization believe it will allow more and more individuals, as consumers and producers, to enjoy the benefits of economic liberalization, competition, and innovation. It's natural to want to see oneself as part of the solution rather than part of the problem, so the space community undoubtedly would like to view itself as an essential tool of globalization for redressing deficiencies and providing solutions for global problems. But general acceptance of this view is not automatic. In fact, there is a risk that the opposite may occur.

The challenge for space development is to continue its role as a key element of globalization without becoming associated with its negative consequences. The same types of organizations that dominate space development—government institutions and large corporations—are seen by critics as orchestrating globalization to serve the wealthy at the expense of the poor. In this view, observations of the Earth from space might be interpreted as a way to exploit natural resources and spy on economic activities in other parts of the world, rather than being seen as an instrument of environmental protection and disaster relief. Satellite communications might be depicted as a tool for extracting information and capital from unsuspecting regions of the world, rather than as a means of bringing information and capital to them. Even incoming information can be pejoratively portrayed as cultural contamination or western propaganda designed to influence national or regional policies and attitudes.

Space technology could be seen by globalization critics as a tool of transnational corporations that exploit workers, of foreign investors who undermine local businesses, or of wealthy (spacefaring) countries that economically take advantage of developing nations. The result, as Robert Reich has noted, could be neo-Luddite controls on technology and onerous trade protection schemes that suppress economic dynamism. Therefore, it's critical that government-supported space development be directed at—and perceived as—seeking solutions for the planet in areas such as disaster relief, environmental monitoring, climate research, medical research, and in the long term, the use of extraterrestrial resources and capabilities for the benefit of Earth.

For at least part of the Cold War era, large public expenditures on space projects were widely perceived in the U.S. as good investments to counter powerful unfriendly forces in the world that could wreak nuclear destruction at any moment. Today, the public's perception of U.S. civil space efforts as a counterweight to unfriendly forces appears much weaker, although this factor is not measured directly by public opinion polls. Certainly the national prestige argument for the space program, in the sense of bolstering the nation's security, has lost much of its impact, since demonstrations of spaceflight prowess are not likely to win many hearts and minds among nations of concern in the world today, and even less so among non-state actors. Similarly, space achievements of other nations are unlikely to prompt a reaction from the United States comparable to the aftermath of the October 1957 Sputnik launch. China's human spaceflight program, for example, fails to stir fears in the west as it follows a path that was tread by the U.S. in the 1960s.

In the current era, the value of government space efforts needs to be measured by a different yardstick that takes into account the multi-polar geopolitical environment and the globalized nature of economics and technology. The space community must recognize the effect of this environment on trade, technology, and leadership in space, and resist the urge to preserve the outdated aspects of institutions, processes, and relationships that insulate it from the evolving big picture.

Looking to the near-term future, it's easy to be pessimistic about global trends in general and the development of space in particular. Globalization, for all the wealth it has produced, can't prevent resource shortages or worldwide economic recession; indeed, when such problems arise, globalization will help them spread faster and wider. Climate change looms like a terminal illness waiting to strike; no one knows when it will come or whether a cure can be found in time. In space development, major new systems can't seem to meet their cost and schedule targets, and if they underperform or fail, a huge investment of resources has yielded little or no payback. The ever-present gremlins of high cost and high risk for getting into space seem no closer to being solved than they did 20 years ago. And the U.S. space industrial base is suffering from a talent shortage as upcoming generations of workers find it to be a far less desirable career track than previous generations did.

What of the future of globalization itself? Will it continue to flourish in the decades to come, or will it end relatively soon, perhaps suddenly, as it did a century ago at the outbreak of World War I? If globalization abruptly halted—due to war, economic collapse, pandemic, catastrophic climate change, or whatever the cause—would the United States continue to be a world leader? A look back at what happened to Great Britain is instructive. Until the start of the war in 1914, Britain was the world's financial center, but by that time the U.S. had become the world's largest national market and had surpassed Britain in industrial output. After 1914, the hardship of the war and the faltering economic recovery of the following decade shifted financial leadership to the U.S., where it has remained ever since. Think about how familiar that sounds today, except a different mix of countries is involved this time. I'll let Jeffry Frieden pick up the story in this excerpt from his book *Global Capitalism: Its Fall and Rise in the Twentieth Century*. As you read this, try substituting the United States for all the references to Britain, and emerging economic powers like China and India for the references to Germany, France, and America. The description becomes eerily familiar.

> Between 1870 and 1913 the size of the British economy well more than doubled; even if one takes into account population growth, British output rose by more than 50 percent per person in those years. Yet the gap between Britain and the rest of the world narrowed continually. British manufacturers were being beaten out of export markets, even out of the British market. The United States and Germany were the world's manufacturing dynamos; the United Kingdom maintained its leadership only in such services as banking, insurance, and shipping. It was no longer a given that the next power plant or railroad built in Africa or eastern Europe would be British; it was just as likely to be German, French, or American. Even in international investment, Continental financial centers—as well as New York—were challenging London's supremacy. It could hardly have been imagined that Britain's enormous industrial lead would last forever,

but the speed of its erosion led many Britons to ask how this had happened . . .

Could either continuation or disruption of globalization shift leadership roles the way it did in the early 20th century? It happened once, so it can happen again. The story of space development in the globalization era, and of U.S. ambitions in this arena, is still being written.

Trying to remain an optimist

What do these seemingly ominous circumstances tell us about expectations for the United States and the world for the rest of the century? Perhaps we shouldn't read too much into them. Extrapolation of recent trends is one of the best tools we have for projecting what's ahead, but it's far from perfect. Events of the previous century played out differently, and in many cases more quickly, than anyone in 1900 would have predicted. The demonstration of powered flight by the Wright brothers in 1903 amazed the public, but was seen by most as a novelty rather than a practical development. Who would have guessed that in three decades there would be transoceanic passenger flights, and in just 66 years—one human lifetime—people would fly to the Moon!

One of the most influential technological phenomena of the 20th century was the proliferation of the automobile. In 1900, there were only 8000 cars registered in the U.S., and few paved roads. The nation had yet to see its first traffic light or gas station (fuel was purchased at hardware stores) and maintenance was almost always performed by the vehicle's owner. No one at that time could have imagined that by the late 1920s, the nation would be manufacturing hundreds of thousands of cars per year. By mid-century, ownership of cars was the dominant factor in determining human settlement patterns in the U.S., and the government was making a huge commitment that would result in the interstate highway system. By the end of the century, 200 million vehicles traversed about four million miles of American roads.

It's safe to say that the general population in 1900 would have considered much of what was accomplished by the early 2000s (economically, scientifically, and technologically, if not in other ways) to be impossible, or even inconceivable. By 2100, people may recall nostalgically our quaint but inefficient use of chemical propulsion for spaceflight, remember sympathetically our inability to cure cancer, and marvel that we ever got anything done using data networks prone to crashes and viruses.

To date, benefits from space from just three applications—communications relay, navigation, and Earth monitoring—are so widespread and diverse that they defy measurement. Should we expect that in the next century, these three will continue to be the only space applications contributing to our global security, commerce, and habitability? It would seem shortsighted to think so. Materials processing in the microgravity environment of space and mining at lunar settlements, to name just two possibilities, are economically impractical today, but in the span of several decades, that can change.

Space development may experience impressive leaps in capability in this century should economics and events drive public and private investment in that direction. By 2100, global society will tap into extraterrestrial material and energy resources for the benefit of Earth, unless the inhabitants of our planet give up completely on space exploration and development. But given the massive infrastructure requirements of a truly spacefaring society, how can we get there from here? The greatest difficulty in building momentum for macro-systems in space lies in defining what kinds of organizations would be willing and able to finance, develop, build, deploy, and operate such systems. The global scope and expense of such projects would indicate some type of multinational consortium having both public and private participants. But this leaves unresolved the details of the appropriate division of efforts and funding responsibilities, as well as how to share profits and losses.

When faced with a similar dilemma in the 17th century—the economic development of far-flung colonies—human ingenuity invented the corporation, a legal entity that allows investors to limit their liability while permitting large amounts of capital to be amassed. As national economies became global, the decades following World War II witnessed the evolutionary development of the multinational corporation. At the same time, public-private hybrid organizations emerged, both within nations and across international borders, and started along their own evolutionary paths.

Steps such as these made possible projects like geostationary satellite constellations, the International Space Station, the Trans-Alaska pipeline, and the English Channel tunnel. The latter two are of particular interest, since they were financed entirely by the private sector. In the mid-1970s, the Trans-Alaska pipeline system took three years to construct at a cost of over $8 billion (about $26 billion in 2008 dollars). The English Channel tunnel, a six-year construction project completed in 1993, cost $13.5 billion (about $20 billion in 2008 dollars).

Attempts to draw direct analogies between space development and terrestrial development are questionable at best. The space experience presents too many challenges not faced in an Earth-bound environment, at least not to the same degree of difficulty. Having said that, there is value in studying the history of other large and influential developments involving massive infrastructure and driven by technology and market forces. There is much to be learned about the dynamism and resilience of modern society. Despite seemingly insurmountable economic barriers and infrastructure demands, a properly motivated society, in time, can achieve what seemed unachievable only a few years earlier. This was demonstrated in numerous success stories of the last century: aviation, the Apollo program, information technology, biomedical advances, and as mentioned earlier, the widespread adoption of the automobile.

The challenge for the 21st century is to successfully execute macro-engineering projects in space that require more off-world infrastructure than anything yet attempted, including the International Space Station. This presumes that some sector (or combination of sectors) makes a long-term commitment to place its resources into space efforts. Of course, it could turn out that the decades ahead see waning interest in space as other areas come to the forefront, such as biotechnology, robotics, advanced transportation (for example, magnetic levitation trains), and revolutionary means of energy generation (such as fusion or hydrogen fuel cells). But the overlap of technologies, combined with the proliferation of space capabilities around the world, should ensure the continuance of space advances.

We may be witnessing a confluence of events that will thrust space activities into the mainstream. Many analysts see the definition of national security as expanding to include economic and environmental security in addition to military and intelligence concerns, and space plays a role in all of these. Influential factors include the evolution of global telecommunications; the quest for solutions to large-scale environmental problems; the growing demand for energy, especially electricity; improvements in technology that will decrease the cost and increase the performance of space systems; the relaxing of market restrictions in many parts the world; and the worldwide proliferation of expertise in space system design, operations, and utilization.

Our potential as a technological and (hopefully) enlightened society at the beginning of the 21st century is much like that of a child just entering school. Children at that stage have no clue about what they will experience or what will be expected of them during the time it will take to grow into adults. Indeed, if they did comprehend what was ahead, they'd undoubtedly run away in terror. Much will be expected of the global community in the search for 21st century solutions. Running away is not an option.

Chapter 2
Searching for a Vision of the Future

The goal of futuring is not to predict the future, but to make it better. —Edward Cornish, founder of the World Future Society

There is a common misperception that "futurism," "futurology," "futures studies," or "futuring," as it is variously called, is aimed at making predictions of what will happen by a particular point in time. (That's what fortune-tellers and astrologers do—badly.) In fact, futurism uses a variety of techniques, such as trend-spotting, expert opinion surveys, scenario generation, and mathematical modeling, to formulate an array of possible futures and make a reasonable assessment of which ones are most probable. This helps our society's decision-makers in a couple of ways. First, it allows for informed speculation on what may be over the horizon. No one likes to be blindsided, and public policy is not well served by knee-jerk reactions to events and circumstances that should have been anticipated. Second, futures studies emphasize that although current trends are strong drivers, there are multiple possible futures, and actions taken today can target or avoid particular outcomes that may be years or decades away.

Of course, even if everyone embraced long-term thinking, there would still be disagreements on goals and priorities. Political partisanship, regional biases, conflicting interpretations of data, and other differences would still exist. Even in the space arena, which may appear to be primarily technology-driven, there are philosophical arguments that have been going on for decades. Since the end of the Apollo era, for example, U.S. civil space efforts have struggled with questions on the respective roles of the government and the private sector, made all the more difficult by the fact that the answers are moving targets. Who should finance, build, and operate space infrastructure elements such as launch systems and space stations? To what extent should the government support research projects that have the potential to produce private-sector revenues? In an era of tight federal budgets, should the government shift as much responsibility and expertise as possible to the private sector, or is this a short-sighted strategy that will undermine the nation's continuing need for large-scale, evolving space capabilities? Can the private sector, at its current level of technical development and operational experience, always be counted on to choose better space investments and technical approaches than the government? If the government is to remain a major player in technology development, should it direct its efforts at incremental advances or revolutionary leaps? Should the government actively encourage and facilitate space commerce, as NASA and other federal agencies are directed by statute to do, or stay out of it and let the private sector find its own way?

There's another question that seekers of the ultimate truth have been asking for a half century, yet it still has not been put to rest. Should public-sector space investment be focused on human exploration or robotic science missions? If one assumes that humanity's destiny is to venture throughout the solar system on a permanent basis, this is absurdly simple to answer: humans should do those things for which they are uniquely qualified, and robots should do everything else, to the extent they are capable of doing so. (Obviously, as both human and robotic space capabilities evolve, the task list for each will change.) But there is no such consensus on humanity's destiny. Some space proponents have asserted that the next industrial revolution is about to begin, and its focus will be in space. Others remain unconvinced and scoff at this pie-in-the-sky notion, claiming that it will never be practical and worth the risk. If it were, it would be obvious to everyone by now, not just to avid science fiction fans and space geeks. After all, we've been in space for half a century, and no one is living on the Moon yet! The difficulty in resolving these basic disagreements is not surprising. We're contemplating a profound and lasting change to the human condition, and such things tend to happen gradually and often imperceptibly to the society that's being affected. Most people today probably contemplate becoming a spacefaring civilization about as much as fish of the late-Devonian period thought about becoming amphibians.

Space exploration is a product of the imagination of many generations, perhaps going back as far as the ancient Greeks and Chinese. The realization of space exploration in our time is a product of the confluence of several key factors in the post-World War II era, including maturing technologies, the ability to amass sufficient resources for the undertaking, and a political environment that has viewed space activities as beneficial to a variety of national interests.

When humanity first took to the oceans, our boats stayed within sight of land—much like our early spaceflights stay in Earth orbit. Millennia passed before people were routinely sailing all over the world's oceans. How long will it be before we're routinely traveling around the solar system? Certainly this evolutionary period will not be measured in millennia, but it may be measured in centuries, or at least in far more decades than we've logged in so far on this endeavor. We've lived in the space age for less than the length of an average human lifetime, so the amount of progress we've already made has come at a pace that is something akin to greased lightning on the timetable of species evolution.

Despite the rapid development of human society and technology in the past century, don't expect a quick or obvious transition to a truly spacefaring species. I doubt that any individual in human prehistory jumped up one day and said, "Hey everybody! We've just entered the Bronze Age!" Even if this did happen, the overwhelming response would have been, "Who cares? You can't eat bronze." Today's world still has a lot of people who are more concerned about finding food than making bronze. But that doesn't mean we shouldn't dream and plan.

Expectations of a baby boomer

When Alan Shepard became the first American in space on May 5, 1961, I was six years old and my attention was fixed on a black-and-white television set. That 15-minute suborbital flight in a tiny capsule doesn't seem like much by today's standards, but it was a significant milestone for the American space program and a turning point for many young observers like me. Shepard's flight, and President John F. Kennedy's proposal three weeks later to send men to the Moon, sent a clear message that human spaceflight had become a reality for America. From that point on, I was hooked, and I grew up watching the rapid development of space capabilities throughout the decade of the 1960s. It was an evolving story that gave people something positive to follow at a time when too many headlines dwelled on unpleasant subjects like civil unrest, illegal drug use, and the Vietnam War. From my youthful perspective at the time, the race to the Moon seemed more like a high-tech sports rivalry than a bitter Cold War struggle. I thought decisions for the space program were made by assembling our top scientists and engineers to figure out the best technologies we could build, and then applying the people and resources needed to go ahead and build them. Life seemed simpler then.

A decade later, the Apollo program was winding down, and the character of the American space effort changed dramatically. Civil space projects fell almost entirely off the president's agenda, congressional space committees were demoted to subcommittees, NASA's civil service employment dropped by more than a third, and hundreds of thousands of aerospace contractor personnel were laid off. Many observers believe that NASA and the space program never really recovered from this loss of direction, deriving momentum instead from large human spaceflight programs (the space shuttle, the International Space Station, and plans to return to the Moon) driven largely by bureaucratic self-preservation. Meanwhile, the societal benefits from space technologies that NASA helped to create became so widespread that they were taken for granted: primary benefits such as satellite communications, navigation, weather forecasting, and other Earth observation capabilities; and secondary (spin-off) benefits such as accelerated development of electronics and computers, advanced materials, and medical technologies.

When I was in elementary school, my sixth-grade class subscribed to a periodical called *Current Events*, which is still published by Weekly Reader Corporation. Directed at sixth to tenth graders, the magazine enlightened us about what was going on in the world, presented at a level we could understand. For their issue on March 23, 1966, the editors decided to do something different: speculate on future events. The theme of the issue was "Middle-Aged YOU in the Year 2000," and included articles on life in the United States in 2000, weather control, and of course, developments in space in an article ambitiously titled "Life in Space Becomes Commonplace." When I received this issue, I immediately decided to save it until 2000 to see how well its vision of the future held up.

An often-repeated truism is that prediction is very difficult, especially if it's about the future. There's a tendency to *over*estimate short-term developments and their consequences and to *under*estimate long-term changes. *Current Events* was looking ahead 34 years, which is less than half an average lifetime in the U.S., so it's a relatively short time period. Overestimation is clearly evident. Contrary to the magazine's prognostications, by 2000 we were not living in domed or underwater cities, driving hovercraft, or linking into electronic highways that drove our cars for us. Computers and automation did become common, but didn't "free us from almost all routine drudgery." The magazine didn't envision the capabilities and implications of the Internet—no one at that time did—but its prophecies were on the right track, observing that "children may do almost all of their research at home. They may be able to dial an electronic library center. Needed information and perhaps even copies of book pages may be available to them in minutes . . ." My favorite lifestyle forecast, however, is this one:

> Electronic problem-solvers and labor-savers are certain to produce one thing in abundance—free time. One of the big philosophical questions now being asked about 2000 A.D. is: "How will people use their long periods of non-work?"

Now there's a problem I wish I was confronted with.

Sadly, the *Current Events* predictions for developments in space by 2000 were far off the mark:

> Many experts believe that by 2000 man will have colonies on Mars, and possibly other planets. Orbiting space cities may be maintained by several nations . . . By 1980, there should be regularly scheduled trips to the moon. A Lunar International Laboratory may be established . . . Space platforms may also be used for launching manned expeditions to other planets. Such missions will require nuclear-powered spaceships, assembled and fueled in a low earth orbit from materials brought up by earth-to-orbit freighters . . . By the year 2000, people may be flying to [space stations] to spend three-week vacations in space. Orbital resort hotels will have swimming pools, fancy restaurants, and all of the comforts of earth . . . Hospitals and universities may also be established in space.

These predictions seem wildly ambitious from our post-2000 perspective, but at the time, the *Current Events* editors were basing their reports on the statements and writings of respected futurists. But would *Current Events* have fared better if they had tuned their crystal ball only half as far into the future? Not necessarily. In 1986, the *CBS Evening News* did a five-part series that forecast just 15 years into the future—to 2001—based on consultations with a similar array of futurists. Among their many inaccurate predictions: "the Russians could land on Mars" and "Americans will work just six hours a day, just 30 hours a week." There's that short work-week prediction again! Maybe it's just wishful thinking.

Prophets or dreamers? Learning from early space futurists

In my view, a Golden Age of Space Futurism began in the years immediately following World War II and lasted for about three decades. This is not to imply that there were no space visionaries prior to the 1950s. In fact, there were many: Konstantin Tsiolkovsky in Russia, Hermann Oberth and Herman Noordung in Europe, and Robert Goddard in the United States, to name a few. But as mid-century passed, there was a perceptible change in the way policy-makers and the public thought about space. In political science terms, this was a shift in "policy image." Spaceflight ceased to be merely science fiction and became something that could actually happen within one's own lifetime. This was a profound attitude adjustment, stimulated by technological advances and media attention, that turned space into a serious public policy concern that was in need of long-term visions.

In this environment, some space futurists managed to make accurate predictions, although we need to forgive them for missing some important details. When Arthur C. Clarke described the concept of communications relays in geostationary orbit (GEO) in an article in the journal *Wireless World* in 1945, he earned his place as a great technological visionary. But he substantially overestimated the length of time it would take to realize his vision: 50 years instead of the 20 years it actually took. This miscalculation was due to imperfect assumptions about the speed of technological development in the post-World War II era. In 1945, the most powerful rocket available, the German V-2, was incapable of carrying payloads into low Earth orbit (100-1000 kilometers altitude), let alone GEO (36,000 kilometers). The transistor hadn't been invented yet, and given the reliability of the electronics of the day, Clarke believed that the GEO platforms would need to be staffed with full-time repair crews.

Clarke was quite a prolific author of both fiction and non-fiction. Among his dozens of books and hundreds of essays, perhaps the most widely known is his book *2001: A Space Odyssey*, which also involved a collaboration with director Stanley Kubrick for a movie version. A pragmatist as well as a visionary, Clarke avoided violating the known laws of physics even when he wrote fiction set in the distant future. His prose was instructive, and also a lot of fun to read. Two of my favorite examples of his non-fiction work deserve mention here.

In 1963, Clarke collected essays he had written during the previous five years and published them under the title *Profiles of the Future*. He covered a variety of topics, such as terrestrial and space transportation, energy production, manufacturing, the potential of the human brain, artificial intelligence, and the difficulties of time travel and invisibility. His most remarkable predictions were those regarding telecommunications. In an essay written prior to the 1962 launch of the pioneering Telstar communications satellite, he accurately depicted direct-to-home satellite television, portable phones with position location capabilities, and fax machines. He also provided a prescient description of the Internet. The daily newspaper, he prophesied, would show up on your TV screen, and would allow you to print out just the articles you wished to read. Further, the same TV screen could give you access to an electronic library of books and documents.

Clarke's clairvoyance did not extend to all areas, however. In the back of the book he placed a "Chart of the Future," which he advised readers not to take too seriously. This was good advice, since the chart predicted nuclear rockets in the 1970s, planetary expeditions in the 1980s, artificial intelligence and fusion power in the 1990s, planetary colonization in the 2000s, weather control in the 2010s, and interstellar probes in the 2020s.

A few years later, in 1968, Clarke produced one of his most memorable non-fiction works, *The Promise of Space*. This is undoubtedly one of the best tutorials ever written for a general audience on rocketry, orbital mechanics, and the solar system. Exercising his predictive powers, Clarke wrote of reusable boosters and aerospace planes, and finished with a discussion of the challenges of traveling to the stars. Although his aim in this book was not to predict when certain capabilities would arrive, Clarke exhibited his optimism when he said, "the exploitation of the foreseeable techniques to their limit could result in truly commercial space transport being in sight by the end of this century. And perhaps fifty years from now [2018], anyone should be able to afford a visit to the Moon at least once in his lifetime."

Clarke was not the only space futurist familiar to the baby boom generation who sometimes overestimated what could be accomplished in a few decades. For example, Dandridge Cole, an aerospace engineer who was also trained in chemistry, physics, and medicine, wrote books, articles, and technical papers that provoked much thought in the late 1950s and early 1960s about possibilities for space development. Cole was a classic out-of-the-box thinker, perhaps best remembered for his 1964 book *Islands in Space*, co-written with science writer Donald Cox, advocating robotic and human journeys to the planetoids (a term he preferred over "asteroids") as the next major space endeavor after the Moon. He envisioned the human spaceflight program as follows:

Phase 1—Exploration of the Earth's Space Environs (1961-65)
Phase 2—Exploration of the Moon and Its Environs (1965-75)
Phase 3—Exploration and Exploitation of the Planetoids and Martian Moons (1975-85)
Phase 4—The Exploration of Mars and Other Solar System Planets (1980-2000)

Cole felt that NASA's thinking at the time represented a go-slow policy. Postponing manned Mars flights to sometime in the 1980s, he believed, would concede the initiation of interplanetary flights to the Russians.

In his 1965 book *Beyond Tomorrow: The Next 50 Years in Space*, Cole acknowledged that the space program at that time was mainly designed to gain technological advantage and win the allegiance of nations in a global power struggle. But from a long-term perspective, he believed that "We could fill books with problems of fundamental importance to the human race which can be solved *only* by spaceflight, *more easily* by spaceflight, or *more probably* by spaceflight."

Cole discussed several emerging developments and took the position that "All of these major breakthroughs are connected in some way with the conquest of space!" He believed spaceflight would stimulate technological breakthroughs toward revolutionary, rather than just evolutionary, advances. He had in mind a diverse mix of research and development areas including energy production (especially nuclear), automation, artificial intelligence, scientific discoveries on the formation of the universe and the basic structure of matter, the improvement of the human species through genetic and mechanical means, and world government. All of these advances he expected "in the next fifty years"—in other words, by 2015. We have made great strides in many of these areas, but we are still far from the levels Cole anticipated.

Cole excited readers with his optimism. Here are some of the key assumptions he made about what would happen within 50 years (and a parenthetical reality check):

- Many advanced space vehicles based on new forms of propulsion will be built. (Actually, we're still using the same basic chemical

rocket concept that has been used since the beginning of spaceflight.)
- Space transport costs will drop by an order of magnitude every seven years for the next two decades. (Much to the chagrin of everyone in the space business, this has not occurred—not even by one order of magnitude in 50 years.)
- Nuclear fission power should be available at much lower prices unless something much better such as fusion power has replaced it. (Obviously, this hasn't happened either.)
- On the question of whether the inner solar system will be colonized during this time period, "the tentative conclusion must be—yes." (The actual conclusion must be—no.)
- Extraterrestrial bases and colonies will become self-sustaining. (Such bases and colonies remain conspicuously absent.)

Without a doubt, Cole's timelines were horrendously off. Does that mean his attempts to paint a picture of the future have no value? Was he simply creating fantasies to inspire science fiction buffs? Not necessarily. Like many of his contemporaries, Cole assumed a pace of development that was far too ambitious. As was the case with Clarke, he was not proposing anything that would defy the known laws of physics, nor was he probing into scientific and technical areas that lacked at least some theoretical or practical background. His work reflects the fact that he was a product of his environment, which at that time made it look like revolutionary developments were just as likely, or even more likely, than evolutionary ones.

Clarke and Cole were tapping into a fascination with space goes back a long way. They had the good fortune of publishing at a time when space had emerged from a techno-geek cult following to capture widespread attention. This started in the early 1950s as post-World War II economic and technological growth provided the means to pursue space activities in the not-too-distant future. Technically grounded, well-thought-out plans for making space travel a reality appeared at this time, the most popular of which came from Wernher von Braun and a team of prominent advocates. Their plan was widely publicized in print media (for example, in eight issues of *Collier's* magazine from 1952 to 1954 that featured the human conquest of space) and on television (most notably in three Walt Disney "Tomorrowland" segments that aired between March 1955 and

December 1957). The steps in the plan, which today is often referred to as the Von Braun paradigm, have been summarized by Smithsonian Institution space historian Roger Launius as follows:

1. Launch and operate Earth-orbiting satellites to learn about the technology and the space environment.
2. Conduct Earth-orbiting flights with human crews to learn how to live and work in space.
3. Develop reusable spacecraft to shuttle back and forth to low Earth orbit.
4. Build permanently inhabited space stations to observe Earth and serve as a launching point to other destinations in space.
5. Explore and settle the Moon.
6. Explore and settle Mars.

To a very substantial extent, this plan has guided U.S. human spaceflight efforts since the beginning of the space age. The Moon race of the 1960s shuffled the order of events—skipping the reusable spacecraft and space stations to more quickly aim for the Moon—but strategic planning since the end of Apollo has focused on getting back on the Von Braun track. NASA has adhered to this plan and presidential administrations have adopted it as their own. Clearly, no other space exploration and development paradigm has been as influential in U.S. policy.

In the 1960s and 70s, futurists of all stripes included space in their calculations. Herman Kahn, well known for his work on nuclear conflict scenarios at the RAND Corporation, is one example. He and some colleagues started the Hudson Institute, a futurist think-tank, in 1961. The Institute's wide-ranging interests during its first two decades included space development. Kahn's popular 1967 book *The Year 2000* (co-written with Anthony Wiener) offered a list of 100 "very likely" technical innovations anticipated for the turn of the 21st century, including these space-related items:

- Direct broadcasts from satellites to home receivers
- Worldwide use of high-altitude cameras for mapping, prospecting, census, land use, and geology
- Recoverable boosters for economic space launching
- Space defense systems

- Artificial moons and other methods for lighting large areas at night
- Permanent manned satellite and lunar installations and interplanetary travel

The first two items appeared ahead of schedule. The next two items were partially addressed by 2000 but still remain in very early stages of development. The last two items are still beyond the horizon, although one could debate whether the International Space Station (ISS) is what Kahn meant by "permanent manned satellite." Most likely, he envisioned something busier and more permanent, as did many of his contemporaries. Nonetheless, by 2001 the ISS began a long stretch of continuous occupation, albeit with a crew that was restricted to just three people for the first several years.

Kahn's book also offered a list of 25 "less likely" possibilities, some of which involved space:

- Major use of rockets for commercial or private transportation (either terrestrial or extraterrestrial)
- Substantial manned lunar or planetary installations
- Planetary engineering
- Modification of the solar system

Commercial payload launch services became common well before 2000, and commercial human spaceflight is taking shape today. However, the other three items on this list are still far down the road. Kahn didn't go into detail in these areas, but it's likely that planetary engineering refers to terraforming, an ambitious endeavor that aims to turn other planets into environments like Earth. But it's hard to say what Kahn had in mind for "modification of the solar system." Perhaps he felt that we should clean up all the asteroidal debris between Mars and Jupiter. I doubt that he envisioned rearranging the planets, although I'm reminded of the Bugs Bunny cartoon in which Marvin the Martian planned to blow up Earth because it was blocking his view of Venus.

Kahn and his Hudson Institute colleagues produced a 1976 book titled *The Next 200 Years* focused on what they called an "Earth-centered" perspective on global development, which assumed limited levels of space exploration and exploitation. However, the authors also introduced a "space-bound" perspective, which would involve

> ... a much more vigorous effort in extraterrestrial activities early in the 21st century, including the eventual establishment of large autonomous colonies in space involved in the processing of raw materials, the production of energy and the manufacture of durable goods—both for indigenous consumption and as exports back to earth or to other solar-system colonies. Such developments would involve substantial migration from earth and could eventually create very new and different patterns of population and product growth, all quite beyond any projections made from a basically earth-centered perspective.

Kahn and the Hudson Institute, at least during this period, viewed space development as a logical and all-but-inevitable step in societal evolution. The same was true of another influential futurist, Princeton physicist Gerard K. O'Neill. Articulating in great depth the space-bound perspective that Kahn's 1976 book only hinted at, O'Neill published *The High Frontier* in the same year, outlining how humanity should begin using the vast resources of space. He believed that the human condition on Earth would be improved by the work of a large population (from tens of thousands to millions of people) residing in space colonies. Extraterrestrial energy and material resources would enable economic growth and prosperity, both on Earth and on space colonies, while preserving Earth's environment. Contrary to what some have alleged, O'Neill's vision was not intended as a resettlement plan to remedy overpopulation, or as a way for the affluent to escape the planet's problems. His ideas were spawned by a search for solutions to societal needs as much as by his interest in the scientific and engineering challenges of space settlement.

O'Neill generally avoided making firm predictions of the schedule for realizing his concept, but believed a "high-orbital facility" could begin within seven to 10 years and be completed in 15 to 25 years. From our perspective today—many years past his 25-year estimate—it's doubtful that even another quarter century will be sufficient to construct the first iteration of O'Neill's vision.

In his 1981 book *2081: A Hopeful View of the Human Future*, O'Neill allowed himself to delve more deeply into the art of prophecy, taking a 100-year view shaped by five drivers of change: computers, automation, communications, energy, and—the factor that sets O'Neill's view apart from many others—space colonies. He speculated that by 2081, "there may be more Americans in space colonies than there are in the United States" and "a voyage of a few days to a space colony will be as commonplace in 2081 as a Caribbean cruise is to us today." This would be enabled, according to the findings of technical workshops reported in the book, because "a rapidly growing industry in space could be established with a total investment comparable to that made in the Alaska Pipeline (which was entirely privately financed) in a time of less than ten years from the start of a substantial development effort." Once such a commitment was made, he believed, there would be "a near certainty of economic payback beginning within five years after investment."

Some have labeled O'Neill's vision as utopian or escapist, but its prophetic inaccuracies stem from overly ambitious timelines rather than a disconnect from reality as these labels imply. O'Neill was well aware of political and economic hurdles, but placed too much faith in the ability of technological progress to overcome them. Nonetheless, he conveyed a compelling message about how space could contribute to serving global societal needs, and in the process he has inspired a substantial number of space professionals and enthusiasts in the generations that have followed.

Knowing how far we have fallen behind the expectations of these and other futurists, we may conclude that space futurism has had a dismal record of guiding policy in any overt way, with the notable exception of the Von Braun paradigm. Even if the majority of these technological and societal propositions had been embraced by policy-makers and the public, one might expect that the resulting policies and programs could have vastly underperformed on their objectives for a variety of reasons, such as unexpected technical hurdles, budget shortfalls, or competition from terrestrial alternatives. Does this mean that space futurism is little better than fantasy, and should not be relied upon to advise public policy?

Before jumping to that conclusion, two important points should be considered. First, space futurists did get some things right, and sometimes even *under*estimated our eventual accomplishments. Futurists helped to inspire efforts to develop geosynchronous communications satellites and to go to the Moon, both of which were major advances that were realized more quickly than even most futurists had imagined. Second, much of what they got wrong was a result of overestimating the pace, rather than the substance, of technological developments. Their poor track record in this area stands out because some of the biggest activities—lunar bases, Mars missions, space mining and manufacturing—have missed (or will miss) their predicted appearance not just by a few years, but by a few decades. Part of this is because some ventures turned out to be technically harder than originally assumed; obvious examples include development of fusion systems for power and propulsion, and human trips to Mars. But the greatest culprits in the extension of timelines have been non-technical: political, economic, and other societal factors. The classic space visionaries typically ignored, downplayed, misinterpreted, or misrepresented these factors so they could keep the focus on technical advancement. Arthur C. Clarke was up-front about this in his book *Profiles of the Future*:

> I am limiting myself to a single aspect of the future—its technology, not the society that will be based upon it . . . I also believe—and hope—that politics and economics will cease to be as important in the future as they have been in the past . . . Politics and economics are concerned with power and wealth, neither of which should be the primary, still less the exclusive, concern of full-grown men.

Purposeful avoidance or inappropriate application of a multidisciplinary approach that includes socio-economic realities is evident in the work of the other futurists. In the examples presented earlier, O'Neill and Cole demonstrated more savvy in this area than many of their contemporaries, though they tended to downplay or misread political signals that challenged the inevitability of technological advances. Kahn, in contrast to the others discussed here, actively sought to integrate socio-economic factors, but did so in ways that often seemed contrived to support his conclusions—a methodology that today we would recognize as "spin."

As noted earlier, there is a tendency in futurism to overestimate the near term and underestimate the long term, although as we've seen in the examples above, sometimes the reverse can be true. Near-term overestimation is a particular problem for space futurists, whose subject matter is by nature very long term, and whose audience includes policy-makers, business people, and others who are looking for information about the immediate future and are seeking results in a few years at most.

Positive visions often take on a grand scale, and space visions are extreme examples of this characteristic. The Golden Age of Space Futurism ended by the early 1980s largely due to reality overtaking ambition. The same realities that ended the Apollo program—changing public policy priorities, budget deficits, a sagging economy, reduced confidence in government programs and technological advances—similarly constricted space futurism at this time. This was magnified by the space shuttle *Challenger* accident in 1986, which quickly extinguished the hype that followed the shuttle's arrival on the scene in 1981. It became evident to all that multi-billion-dollar industries in microgravity materials processing and space manufacturing were not on the immediate horizon after all, and the research leading to these activities would take far longer than many advocates had projected. In fact, it would take nearly three decades to build the orbiting research lab (the ISS) that would allow the commencement of basic experiments.

Up to the mid-1980s, the Apollo experience had led many observers to believe that the near-term delivery of awe-inspiring space developments should be the norm, or even the requirement, for the nation's space efforts. Once this view was shattered, futurists in general began to distance themselves from space visions, and those inside the space community found it more difficult to play a constructive role in public policy. That doesn't mean they stopped trying.

National expert panels attempt to plot the future

The space community—researchers, entrepreneurs, advocates, and the builders and operators of space systems—has continued to hope for a window of opportunity to move space back into a prominent place on society's agenda, where it hopefully would stay for the long term. This could be enabled by technical advances, cost reductions, policy shifts, or triggering events. Although the community has tried to remain prepared should such a window open, an environment akin to that which brought about Apollo has remained elusive. But even during the period of the mid-1980s to early 1990s, when bold new space initiatives were a very hard sell, preparations included a series of expert panels reporting to the highest levels of the U.S. government.

The National Commission on Space (NCOS) was mandated by the Congress in 1985 to do a one-year study to formulate a plan for the next 50 years of U.S. civil space activity. Chaired by former NASA Administrator Thomas Paine, the 15-member group included Princeton futurist Gerard

K. O'Neill, Apollo astronaut Neil Armstrong, and famed aviator Chuck Yeager. Although chartered by the Congress, the Commission's members were appointed by President Ronald Reagan, and the group reported to NASA, which provided its funding. As one would expect of an agency that had much to gain or lose, NASA tried (mostly unsuccessfully) to influence the content of the report. For its part, the White House didn't appreciate being told by the Congress to seek outside help in setting its space policy.

When the NCOS was established, it was instructed, to the chagrin of many of its members, to assume that NASA's space station program would be carried out as planned, thus prohibiting the Commission from debating the station's merits or eventual configuration. This assumption was unduly restrictive, considering that the already controversial station was still in its preliminary design phase. In fact, from the time of the NCOS study to the space station's completion 25 years later, the design underwent five major and countless minor changes. One would think that the Commission should have had some say in space station design concepts and mission planning, since the station would be a vital piece of infrastructure throughout most of the 50-year timeframe of the study.

The Commission's report outlined major steps in infrastructure development and recommended concurrent efforts in the civil and commercial sectors. However, there was almost no follow-up from the Reagan administration. A primary reason was bad timing—the Commission's report came out at around the same time as the far-more-newsworthy Rogers Commission report on the space shuttle *Challenger* accident, and soon after that the administration became embroiled in the Iran-Contra controversy. The only significant policy directive to emerge from the Oval Office that was related to the report's recommendations came nearly two years later (issued in January 1988, publicly released the following month) and was deemed too little, too late by most analysts, since the Reagan administration would be leaving office less than a year later. It included only one paragraph that referred to human exploration beyond Earth orbit, and the Moon and Mars were not specifically mentioned. The directive recognized that there should be a "long-range goal of expanding human presence and activity beyond Earth orbit into the solar system," but it put off any serious commitment by simply directing NASA to "begin the systematic development of technologies necessary to enable and support a range of future manned missions." Such development would be "oriented toward a

Presidential decision on a focused program of manned exploration of the solar system."

About a year after the NCOS report, NASA issued "Leadership and America's Future in Space" (also known as the Ride report because it was chaired by astronaut Sally Ride), which can be viewed as NASA's internal response to NCOS. It offered four scenarios, or "leadership initiatives" as the report called them: Mission to Planet Earth (in other words, Earth science), Exploration of the Solar System (in this case referring to robotic missions only), Outpost on the Moon, and Humans to Mars. The report revealed the study's "ground rules set forward at the outset of this study," the first of which was the following:

> The initiatives should be considered *in addition* to currently planned NASA programs. They were not judged against, nor would they supplant, existing programs. [Emphasis in original]

This not only takes the space station as a given, but an assortment of other programs as well. NASA budget increases were expected, the magnitude of which would depend on the mix of programs chosen from the study's four initiatives. The Mars initiative apparently was predestined, since the following was also listed as one of the study's ground rules:

> The Humans to Mars initiative should be assumed to be an American venture. It was beyond the scope of this work to consider joint U.S./Soviet human exploration.

This assumption is a demonstration of the continued dominance of the Von Braun paradigm. Like many other civil space planning documents before and since, Humans to Mars was treated as the ultimate objective, justified only by general statements about scientific achievement, national pride, and inspiration of youth.

Two years after the Ride report, a White House-chartered panel of experts called the Synthesis Group sought to provide a more detailed assessment of space futures, at least with regard to lunar and Mars mission scenarios. As in the Ride report, four scenarios were presented. But unlike Ride's approach, all four described different ways of doing the same thing: human missions to the Moon and Mars. The Synthesis Group didn't have the option of excluded Mars, perhaps to suggest another path for human space efforts such as devoting the next two or more decades solely to exploration of the Moon. Had the Group been instructed to study the questions of "what, why, and when," instead of just "how," its findings may have been quite different. Evidently, these questions were thought to have been answered before the Synthesis Group began its work. The Group's 1991 report included a sales pitch to convince the reader that the rationale for Moon and Mars journeys had already been validated.

The Synthesis Group's approach was driven by comments made by President George H.W. Bush in his July 20, 1989 Apollo anniversary speech in which he stated that the U.S. would go back to the Moon and on to Mars. Once again, the Von Braun paradigm dominated. Bush initially submitted sizable requests for start-up funding for the Space Exploration Initiative (SEI) and allowed Vice President Dan Quayle and the National Space Council to devote the bulk of their efforts to promoting the concept. The idea was poorly received in Congress, which granted only token appropriations for preliminary studies. Even that meager funding dried up by 1993.

Unlike the National Commission on Space and the Ride study, the Synthesis Group did not report directly to the NASA leadership because it was chartered by the White House National Space Council to seek out ideas other than NASA's. The Synthesis Group was formed in the aftermath of an internal 90-day study at NASA, which offered the space agency's response to Bush's Apollo anniversary speech and supplied the initial conceptualization of what became SEI. NASA's 90-day study, delivered to the White House in November 1989, described in some detail robotic precursor missions to Mars, while the Moon was treated as merely a way-station on the road to Mars. Vice President Quayle was dissatisfied with NASA's recommendations and sought wider participation in planning the ambitious programs. The Vice President was interested in faster, more cost-effective methods of performing the missions, so the Synthesis Group became the clearinghouse for suggestions from around the space community on technical approaches to sending humans to the Moon and Mars.

Around the same time that the Synthesis Group was active, a NASA-sponsored panel known as the Augustine Committee (named for its chairman Norm Augustine of Martin Marietta Corp.) was given a broad mandate to answer the "what, why, and when" questions, and to an extent the "how" questions, but was given only 120 days to do so. Contrary to the expectations of some, the Augustine Committee's 1990 report contained nothing significantly disruptive to the status quo, such as splitting the agency into separate research and operations entities, or radically changing the way field centers are managed. Instead, it called for higher priority for space science missions, the development of a new heavy-lift launch vehicle, and a funding increase of 10 percent per year over the decade of the 1990s. The U.S. Congress either rejected or ignored these recommendations just a few short months later, denying the level of support that the Bush administration requested for fiscal year 1992 and raiding space science programs to pay for the space station. The only apparent response to the Augustine panel's recommendations was some token reshuffling of management duties at NASA headquarters.

In general, the impact of advisory reports such as those described above can be affected by the circumstances under which the report is delivered, through accident or by design. As mentioned, the 1986 NCOS report was released within a couple of weeks of the Rogers Commission report on the space shuttle *Challenger* accident, which took the spotlight and set the tone for the community to retrenchment and self-analysis rather than ambition and expansion. The Ride report came out in 1987 as the Congress went into its August recess, at which time NASA's top agenda item was getting its budget, and especially the space station, through the last critical phase of the annual budget cycle. The Augustine report appeared just before Christmas 1990 as Congress was ordering NASA to once again shrink the space station design. And the Synthesis Group report, which made minimal mention of the space station, arrived in spring 1991, just before the House Appropriations Committee recommended canceling the station, possibly deriving support for this position from the Group's cursory treatment of the station.

All of these reports had little impact due to prevailing circumstances. Some of these circumstances were beyond the control of the study panels and their sponsors, but many were not. While they may have pleased their sponsors when they were released, these reports were not the vehicles by which a constituency for space goals could be built.

The studies reviewed here had as their goal the reshaping of space policy, but they failed because they behaved as though they were operating in an atmosphere of enthusiasm when in fact it was one of criticism. By this time, the policy community had been taking an incremental approach to civil space issues for many years, and was not prepared to dramatically change course and welcome the comparatively bold advice of the expert panels.

Compounding the problem of a poor political environment, the experts had skipped important steps. "How to" reports should follow definitive "what" and "why" directives. Only the National Commission on Space seems to have taken this into consideration, but the group was disbanded before adequate support among policy-makers had taken shape.

The "blue-ribbon" panels made assumptions, either expressed or implied, that policy-makers, and many of their constituents, later refused to accept. This resulted in attempts to answer some of the wrong questions. For example, devising methods to put a crew on the surface of Mars for a 600-day stay, as the Synthesis Group did, is an interesting exercise, but it has no political value until someone figures out what that crew will do for those 600 days that is important to national or global interests. Unsubstantiated claims of spin-off benefits (for example, that the activity will boost education and the economy) are met with great skepticism when they are professed by panels made up largely of engineers, astronauts, and aerospace company executives who have a vested interest in selling such programs. Policy-makers need to be convinced of these benefits, as do constituents who seek a more substantial rationale than "It's our destiny."

For most of the time since the mid-1980s/early 1990s string of high-level goal-setting studies, there has been a lack of such activity at the top levels of the U.S. government. It was replaced by a series of commissions identifying problems with the government's space organization and management, its space system procurement practices, the U.S. private sector's declining space industrial base, and the U.S. education system's diminishing output of qualified scientists and engineers. Instead of setting bold new goals for the long-term future, the space community entered an extended period of hand-wringing over challenges that threatened the long-term sustainability of its missions. There was, however, one exception in the civil space arena, driven by the loss of the space shuttle *Columbia* in February 2003.

At the time of *Columbia*'s tragic flight, an interagency process led by the National Security Council and the office of the president's science advisor was putting the finishing touches on a new national space transportation policy. The shuttle accident forced them to go back to the drawing board, and to consider more than just space transportation issues. The accident had raised questions about the viability of continuing the shuttle program—which many at that time still hoped would continue beyond 2020—and indeed the very purpose of human spaceflight. This time, there would be no publicly announced expert panel to deliberate on the next steps. Instead, the remainder of 2003 was devoted to secret interagency consideration of new goals for the civil space program that would be revealed in January 2004. But the new goals proved to be remarkably familiar.

The Emperor's New Spacesuit

In the early 19th century, Han Christian Andersen wrote a short story called "The Emperor's New Clothes" in which a vain leader was swindled by a couple of con men who promised him a magnificent new wardrobe possessing the curious magical property that it would be invisible to people who were stupid or unfit for their post. (Interesting concept—a "stupid or unfit" detector could be very useful in so many situations.)

The two con men took the money and materials they were given for their task and pretended to work on production of the nonexistent apparel. The emperor's officials, and eventually the emperor himself, visited the workshop to inspect the goods, but none were willing to admit they couldn't see the supposedly magnificent clothing for fear of being labeled stupid or unfit for their post. When the emperor later paraded through the streets in his skivvies, his subjects—with the exception of one child—were also unwilling to admit what their eyes didn't behold.

So what does this have to do with planning the future of the American space program? Are the nation's space exploration efforts a con job that will never produce anything of substance? Maybe an extreme cynic would say so, but I'm a cautious optimist so I won't go that far. Rather, the point is that the taxpaying public—and possibly President George W. Bush himself—were sold the idea of a new "vision" for space exploration that in fact was neither new nor visionary. It was simply a reaffirmation of the half-century-old Von Braun paradigm.

When the president unveiled his space exploration initiative at NASA Headquarters on January 14, 2004, I eagerly watched the webcast of his speech. There had been rumors for weeks that some kind of announcement would be made soon, but I was skeptical. The president had little to gain and much to lose by taking this step. It could be seen as an election year gimmick to make him look visionary. The timing of the announcement could also be questioned by those who saw such an expensive, long-term commitment as inappropriate at a time when the nation was militarily engaged on two fronts and experiencing record budget deficits. Opponents of the administration's defense policies could view this action as providing a non-military avenue for development of space technology (such as nuclear power and propulsion) for military purposes. And advocates of existing space projects could perceive it as a scheme for squeezing the space shuttle, space station, Earth science, and aeronautics out of NASA's budget. But what did the president have to gain in the (potentially) five years he had remaining in the White House? Not much, except the gratitude of those who were waiting for him to fill the void in policy and planning related to NASA's future in the wake of the February 2003 loss of space shuttle *Columbia*. Certainly, the U.S. space industrial base needed a boost to its sagging ability to attract the best and brightest young minds. But anything produced by the exploration effort wouldn't appear until long after Bush had left office. Given the risks and the limited payoff, I was betting that he wouldn't announce anything bold.

Actually, he did announce something bold, and he did face all of the criticisms mentioned above. But bold is not the same as visionary. After listening to President Bush's speech and reviewing the related documents posted on the White House website the same day, my reaction was, "That's it?" As an avid follower of space futurist literature spanning decades, I suppose I was bound to be disappointed. But let's face it, journeying "to the Moon and on to Mars" is something we've heard many times before—including from the elder President Bush fifteen years earlier—and the concept has never been able to gain much political or public support.

This time was no exception. Some outside the space community responded to the proposal with disinterest, and some were openly hostile, seeing it as a waste of resources at precisely the wrong time. Other reactions could be described as incredulous, as evidenced by these excerpts from a *Washington Post* editorial published the day after the president's announcement:

> ... The nation faces a yawning budget deficit, educational and health needs, and an international terrorist threat. That makes this an odd moment to embark on a dispensable project of great expense . . . More serious than the price, though, is the absence of clear arguments for the project . . . the need for space travel just for space travel's sake is questionable. There must be concrete scientific reasons to set up a permanent colony on the moon . . .

President Bush's policy, officially known as National Security Presidential Directive (NSPD) 31, "U.S. Space Exploration Policy," stated that its "fundamental goal" was "to advance U.S. scientific, security, and economic interests through a robust space exploration program." This is very general top-level guidance, so we need to look for a bit more clarity in the supporting policy objectives:

- Implement a sustained and affordable human and robotic program to explore the solar system and beyond.
- Extend human presence across the solar system, starting with a human return to the Moon by the year 2020, in preparation for human exploration of Mars and other destinations.
- Develop the innovative technologies, knowledge, and infrastructures both to explore and to support decisions about the destinations for human exploration.
- Promote international and commercial participation in exploration to further U.S. scientific, security, and economic interests.

That tells us a little more about the overall direction, but to discover what the president actually wanted NASA to do, we need to go to the next level down—the policy's implementation guidelines. For the sake of brevity and clarity, I'll paraphrase them here:

- Focus the space shuttle program on completing the International Space Station (ISS) by the end of the decade, and then retire the shuttle.
- Focus U.S. research aboard the ISS on supporting space exploration goals (in other words, phase out U.S. research that is not directly linked to long-duration human spaceflight).
- Begin robotic missions to the Moon by 2008 and human expeditions between 2015 and 2020, with an eye toward later human missions to Mars and other solar system destinations.
- Conduct robotic exploration of Mars and other solar system bodies, as well as telescopic searches for Earth-like planets around other stars.
- Develop the systems that will be needed for long-duration missions throughout the solar system, such as power, propulsion, and life support.
- Conduct human missions to Mars when the knowledge and experience to do so has been achieved.

- Develop a new crew exploration vehicle by 2014 and a separate cargo vehicle as soon as practical and affordable.
- Pursue commercial and international participation in supporting exploration missions.

By any measure, that's a challenging civil space agenda that would keep NASA busy for a long time. It could provide a needed stimulus to the U.S. aerospace industry and could produce some beneficial technology spinoffs. Space professionals and advocates quickly heralded the plan as an answer to their prayers since the end of Apollo—presidential leadership toward real space goals! How can any lifelong space enthusiast fail to see this as wonderful news?

Many supporters of the plan bought into the notion that this is the space goal we'd all been awaiting for 30 years, and if we didn't embrace it, we would lose the opportunity to advance space exploration for at least a generation. A popular argument was that we've been stuck going around and around uselessly in low Earth orbit, and now we'll be able to break out of that rut and do real space exploration. Columnist Charles Krauthammer endorsed this view in a January 16, 2004 commentary:

> NASA gave us the glory of Apollo, then spent the next three decades twirling around in space in low Earth orbit studying zero-G nausea. It's crazy, and it might have gone on forever had it not been for the *Columbia* tragedy. *Columbia* made painfully clear what some of us have been saying for years: It is not only pointless to continue orbiting endlessly around the Earth; it is ridiculously expensive and indefensibly risky.

Krauthammer and others who take this position illustrate the difficulty—one might even say the curse—of trying to map the future of spaceflight in the long shadow of Apollo. Sending humans to destinations at ever-increasing distances from Earth is assumed to be the essential element of any plan. Anything less is seen as wasteful, or at least not visionary. The space shuttle and space station are ridiculed for going around and around in circles (which is what NASCAR racers do all the time, and millions of people love it!). The potential benefits of such activities are downplayed or completely ignored. Krauthammer calls low

Earth orbit activities "ridiculously expensive and indefensibly risky," yet the destination-driven approach he advocates would be far more so.

Basing our space exploration and development planning on the assumption of a destination-driven human spaceflight requirement can lead to a dangerous path in which unfulfilled promises and lack of relevance to a broader constituency will turn political will against space activities for an extended period of time. What is needed instead is a capabilities-driven approach, which will be explored further in later chapters. Basically, it begins by asking the question, "What do we want to achieve, both in expanding our horizons and in bringing benefits to Earth?" rather than "Where should we send humans next?"

Almost immediately after taking office, the Barack Obama administration recognized the unsustainable path that the nation's civil space program was on, and established yet another presidentially appointed panel to develop viable alternatives. The panel's charter was limited to the human spaceflight program and its duration to just 90 days, so there was no expectation that it would re-write the conceptual underpinnings or strategic priorities of the U.S. space enterprise. But the committee, under the chairmanship of the same Norm Augustine who led a similar effort 19 years earlier, did try to reorient thinking away from the destination-driven strategies and toward a capabilities-driven approach:

> Planning for a human spaceflight program should begin with a choice about its goals—rather than a choice of possible destinations. Destinations should derive from goals, and alternative architectures may be weighed against those goals. There is now a strong consensus in the United States that the next step in human spaceflight is to travel beyond low-Earth orbit. This should carry important benefits to society, including: driving technological innovation; developing commercial industries and important national capabilities; and contributing to our expertise in further exploration . . . Crucially, human spaceflight objectives should broadly align with key national objectives.

This critically important message from the 2009 Augustine Committee was largely lost in the noise. Far more attention was given to the panel's proposed options for reconfiguring NASA's near-term human spaceflight programs through alternative system architectures, stretched schedules, and of course, additional money.

If the community of stakeholders—in government, industry, and advocacy organizations—presses ahead without justifying space projects by their potential to create new capabilities or bring benefits to Earth, they would be acting like the emperor's subjects in Andersen's tale by refusing to acknowledge the lack of substance and the political and economic weakness of the projects, lest they appear unfit for their post (that is, disloyal to the cause).

A space vision should consider a wide range of activities and recognize that exploration and development should be pursued concurrently and in a complementary fashion. It should also recognize the interdependence of public and private sector efforts. And it should have some hope of being realized, meaning that technical challenges, political feasibility, and affordability cannot be dismissed just because they get in the way of the dream.

Today's self-appointed space visionaries too often allow their conceptualization of the future to get locked into a single large project, such as a human mission to Mars. Similarly, policy-makers continue to address civil and commercial space much the way they did in the Cold War era: flagship human spaceflight programs for national prestige and other foreign policy purposes, and job-creating projects to be distributed around the country. Next, we'll explore these tendencies further and examine how incentives for short-term thinking and attachment to traditional ways of doing things have compounded the challenges of making good choices for the future.

Chapter 3
Muddling Through with a Short-Term View

A politician looks forward only to the next election. A statesman looks forward to the next generation.—Thomas Jefferson

As noted in Chapter 1, the current era of globalization began in the aftermath of World War II. An impressive array of important decisions with long-term technological and economic implications ensued over the next couple of decades. Among them were: the Bretton Woods agreements to establish the post-war international financial system (which in fact started to take shape even before the war was over); the Marshall Plan, which accelerated Europe's recovery from the war; the G.I. Bill, which helped returning U.S. military personnel to continue their education; the creation of organizations such as the National Science Foundation (1950) and NASA (1958) to facilitate government investment in science and engineering research; the U.S. interstate highway system, which became a boon to the nation's economic development; the National Defense Education Act of 1958, which boosted higher education in science, math, engineering, and languages in response to Sputnik and other Cold War concerns; and the recognition that environmental protection required immediate and ongoing attention, resulting in the National Environmental Protection Act, the Clean Air Act Amendments, the creation of the Environmental Protection Agency in 1970, and the Federal Water Pollution Control Act of 1972 (amended by the Clean Water Act of 1977). All of these actions shared certain characteristics: significant up-front investment by the public sector, long lead-times before observable results could be expected to appear, and the desire to steer societal behavior and investment in positive directions through public policy and law.

Certainly, not every major public policy action taken during the first three decades after the war produced the long-term results that were intended. For example, the highly successful recycling program of the war years was abandoned when the conflict ended, contributing to waste management problems across the country and the need to recreate new recycling programs decades later. The so-called War on Poverty (1964), War on Drugs (1969), and War on Cancer (1971) did not eliminate or dramatically reduce poverty, drug abuse, or cancer, although efforts are ongoing. Also, large government investments in commercial nuclear power were supposed to result in electricity that was "too cheap to meter," but obviously that didn't happen because we're still paying electricity bills based on meter readings. All of these actions seemed like a good idea at the time.

There will always be programs that don't pan out, so some level of failure should be accepted. What should not be tolerated is unwillingness to try forward-looking ideas, or to give them a fair chance to prove themselves. Unfortunately, the unwillingness to take risks on creative solutions started to grow in the late 1970s and is still with us today. What happened, and why has it persisted? I believe it can be explained this way: the nation—its leaders and population alike—became unwilling to accept the three characteristics noted above (significant up-front investment, long lead-times, and the desire to steer societal action through government mechanisms).

Large up-front investments of public dollars have always invited close scrutiny from policy-makers, as one would expect, and this intensifies during periods of persistent deficit spending. By the late 1970s, the country had seen a decade of continuous federal budget deficits and a growing national debt. Although the numbers were small by today's standards, they were enough to put a damper on spending at the time. There was plenty of blame to go around, and people could pick their favorite culprit depending on their political leanings: the ballooning cost of the Great Society social programs put in place in the 1960s; the cost of the Vietnam War; the rising cost of energy, for which foreign sources of oil were believed to be mostly responsible; and the increasing competitiveness of the world market as the economies of Europe and Japan completed their recovery from the devastation of World War II. All of these circumstances were important contributing factors, as were successive presidents and Congresses who avoided making difficult decisions about spending and taxation, each hoping that the other branch of government would take the responsibility and suffer the backlash. The longer that deficit spending continued and the national debt grew, the more it foreclosed options for new programs to build infrastructure or conduct research. By 2008, the annual interest payments on the national debt were approximately thirteen times NASA's budget.

Long lead-times—years, or even decades, between initial investments and the appearance of noticeable results—are problematic for elected officials who want to see positive outcomes while they're still in office, or at least be remembered for their role in producing the eventual benefits. Investments in long-term projects such as space research and development may or may not yield visible benefits adequate to justify the cost. For those that do, the benefits are likely to be not only far in the future, but also widely dispersed, disassociated from their origin, and impossible to measure. Such a situation provides little incentive for political actors to invest their resources in fighting for these projects, and the taxpayers' resources in paying for them. As for the taxpayers, they don't like to be kept waiting for their return on investment, and when it comes, they want to see its effects in their neighborhood.

The desirability of U.S. government efforts to steer societal behavior and investment became questionable as citizens lost faith in the capability and credibility of federal agencies and institutions. The influence of the Vietnam War and the Watergate scandal on this loss of faith has been widely recognized, and certainly the "energy crisis" of the 1970s didn't help matters, as the population watched inflation and interest rates rise steeply while U.S. leaders appeared helpless in their attempts to craft and implement meaningful remedies.

Futurism vs. pessimism

The societal evolution needed to tackle today's and tomorrow's national and global issues may require generational turnover among decision-makers and the workforce because it involves changing established habits, procedures, institutions, and infrastructures. These mechanisms, once entrenched, are hard to replace or significantly alter because large communities of people become comfortable with them and depend on them. They dominate the available funding for extended periods of time, and spawn networks of bureaucracies and contractors. Even when there is agreement about replacing them, it takes substantial (and preferably stable) funding for 20 to 30 years to go from the conceptualization of new approaches to full implementation of new systems.

Clearly, decision-makers need all the help they can get from perceptive analysts who engage in long-term thinking. Unfortunately, they are either not getting or not using this help.

There is no shortage of organizations and analysts—for example, think tanks, academic groups, and consultants—engaged in the study of possible long-term futures and the actions needed in the near term to get there smoothly. Some represent themselves as "consulting futurists," such as the dozens that are listed on the website of the World Future Society. Others shun labels like "futurist" but nonetheless advertise themselves to private sector and government customers as astute strategic planners, including the countless consultancies that surround centers of political, economic, and technological power, such as Washington's "beltway bandits." However, many decision-making systems in our society tend to underutilize the futurist perspective, or simply dismiss it as too far out to be relevant.

Futurists are not immune to shifting fashions. This should not be surprising because they need to make a living just like everyone else, and they can't do that by trying to sell a product that has few interested customers. The exploration and development of space is a prime example of an important futurist topic that has fallen out of fashion. In the 1960s and 70s, when new space technologies were being introduced at a rapid pace and Project Apollo was on its way to the Moon, every forward-looking investigator felt compelled to recognize a substantial role for these technologies and the movement outward into the solar system, as demonstrated by those profiled in Chapter 2. Since that time, space has been largely ignored in discussions generated outside of the space community. Even some organizations that once gave substantial attention to space have reduced their efforts in this area. There have been a few exceptions, and controversial topics such as space weapons continue to garner plenty of attention from think tanks and pundits, but in general there has been movement away from space topics in favor of other issues deemed more timely (such as globalization, homeland security, and climate change) with little or no recognition of the contribution that space is making, or can make, in these areas.

Advances in space applications and human spaceflight in the 1960s made it logical for optimists to portray space as an important part of humanity's future, and to expect that the rapid development of space technology would continue. But the seeds of pessimism were already being sown in the U.S. by several influential developments. The escalation of the Vietnam War and racial tensions spawned protests and violent civil unrest, resulting in falling public confidence in the federal government's competence. At the same time, there was increasing awareness of environmental problems after Rachel Carson's 1962 book *Silent Spring*. History shows that this was not the first time a society raised concerns over war, ethnic conflict, or environmental degradation. However, the significance for the space age is that although the space program was at its most popular and was producing historic achievements, many in the 1960s began to see technology as part of the problem rather than part of the solution. Combined with the troubled economy of the 1970s, this contributed to the sharp downturn in the priority and pace of space development.

Space futurism remained prominent in the 1970s, with space professionals and enthusiastic young people riding the momentum of Apollo-era accomplishments and generating remarkable futurist scenarios like O'Neill's space settlement concept noted earlier. The decade-long slump in activity and public interest, particularly in human spaceflight, seemed like a cyclical slowdown that would turn around dramatically once NASA's space shuttle became operational in the 1980s. But at the same time there began an outpouring of cautionary literature that continues today, as indicated in Chapter 1. Discussions of societal and global problems of increasing magnitude made spaceflight look less relevant. Inflated perceptions of the cost of the space program caused some to see it as a drain on resources that would be better spent elsewhere. (When asked in polls to guess NASA's percentage of the federal budget, Americans tend to believe it's in the range of 20 to 25 percent, when in fact it has been less than one percent since the end of Apollo.) Too few saw that space applications could be valuable tools in the search for solutions.

The 1980s brought new possibilities for space, but also a great deal of hype aimed at investors and enthusiasts in search of instant gratification. The space shuttle system failed to live up to its promises of cheap, frequent access to orbit, and then failed catastrophically in the 1986 *Challenger* accident. New commercial space efforts—primarily launch services, remote sensing, and microgravity materials processing—didn't quickly produce profitable multi-billion-dollar industries as many proponents had hoped, and in fact are still dependent on government subsidies today. (More on these topics in Chapter 5.) The Apollo generation was having trouble convincing the next generation that a career in the space community would be exciting, lucrative, and professionally fulfilling.

The end of the Cold War at the beginning of the 1990s shut down the already sagging superpower rivalry in space, removing the last vestiges of what had once been the civil space program's greatest driver. NASA spent the rest of the decade as the nation's poster child for the Clinton administration's Reinventing Government effort (that is, downsizing) and adopted the mantra of "faster, better, cheaper." This was not an effective strategy for the space agency. The joke at the time was, "Faster, better, cheaper—pick any two." Even if this strategy had worked as advertised, it was little more than an acquisition philosophy, not the long-term strategic vision for the future in space that was needed and is still lacking.

As we saw in Chapter 2, the search for a vision continued sporadically during these post-Apollo decades, but without success. By the time the second President Bush announced his space exploration plan in January 2004, the United States had spent 35 years debating at the highest levels on what the nation should be doing in space in the post-Apollo era. Societal developments that changed public policy priorities and contributed to the nation's inability to produce a strategic vision included persistent federal budget deficits, economic problems, and reduced confidence in government and technology, as noted earlier. It's illustrative to look at how these circumstances undermined the Space Exploration Initiative (SEI), the Moon-Mars program of the first President Bush, which was introduced briefly in the previous chapter. This will also serve as a transition to our consideration of how systemic problems combine with societal developments to work against such innovations.

The primary reason for the failure of SEI at the beginning of the 1990s was that virtually no one outside the immediate stakeholders in the space community was willing to commit to a decades-long project of such magnitude in a time of record federal budget deficits. The 90-day study that NASA performed as a follow-up to the president's announcement was widely viewed as a long wish-list of programs that had been waiting for a window of opportunity for many years. Thirty-year cost projections in the range of $300 billion to $400 billion were enough to turn SEI from a bold vision into an embarrassment for the president.

Another reason why SEI was short-lived was the president's apparent loss of enthusiasm in the face of opposition to the program. After the initial rhetoric, he never again intervened on behalf of the program. Perhaps his support for the venture had been weak from the beginning, or perhaps he decided it wasn't worth investing time and effort in something which may not show results until 30 years later. In any case, the work of promoting SEI was left to the National Space Council, headed by Vice President Dan Quayle, but its actions proved to be divisive rather than unifying. The Council's immediate rejection of NASA's 90-day study, and its creation of a study panel (the Synthesis Group) to solicit cheaper alternatives from outside the space agency, alienated SEI proponents inside NASA. At the same time, NASA Administrator and former astronaut Richard Truly, who wanted the agency to stay focused on the shuttle and space station, openly expressed his disdain for SEI, further complicating the agency's relationship with the Council. This friction between Truly and the Council eventually led to Truly's departure as Administrator in early 1992.

Meanwhile, the National Space Council's relationship with Congress was characterized by poor communication and little cooperation. Concerned congress-members came to view the Council, which the legislature had eagerly endorsed just a short time earlier, as a means of wresting control of space policy from the Congress and NASA and consolidating it in the White House. Members began to call for more access to the inner workings of the Council, and Senate confirmation of its executive secretary. In short, presidential neglect and conflict between the branches insured that SEI would not build the coalition it needed to succeed.

This experience highlights the dysfunctional relationships that can form between the White House and its agencies, and between the White House and Congress. We've come to expect this kind of conflict. We see it in the news on a daily basis, especially when the White House and Congress are controlled by opposing parties. The history of the origin of the U.S. Constitution tells us that tension between the branches of government was designed into the system, and it serves a useful purpose. But rather than just accepting it every time it appears, we should question what it means for the long-term national interest. If one of the branches demonstrates a failure of leadership in farsighted strategic planning, why doesn't the rival branch always pick up the leadership mantle? Can this problem be blamed simply on a lack of political courage among specific individuals, or, given the persistence of the problem, is there something intrinsic to the system that discourages long-term thinking and actions? A July 2008 Report to Congress titled "Leadership, Organization, and Management for National Security Space" had a key finding that applies well beyond its intended subject, to all of the nation's long-range strategic interests: "No one's in charge." Can we afford as a nation to let this continue?

Systemic barriers to innovation

The defining task of the nation's decision-makers, I believe, is to shape the future in the national interest, with due consideration of global implications. In a representative democracy defined by the rule of law and the strength of its institutions, this means more than just approving and funding programs that ensure the strength of federal agencies; it is a responsibility to holistically address the needs and enable the standard of living of an entire nation. A list of national interests for a country like the United States includes, at a minimum: preservation of fundamental values and institutions; a stable and secure international environment, free of major threats; a healthy and growing economy; and healthy international alliances linked by a fair and open trading system. Quite a tall order, and one that demands long-term strategic thinking. Decision-making at this level affects the whole world through global relationships that are a mix of cooperation and competition.

Business decision-makers, who also need to shape the future, have a much narrower scope. Their constituency is not the whole of society but rather their own stockholders, whose overarching demand is the sustainable generation of profit, and their customers, who must be kept happy and hopefully will increase in number to enable the sustainable generation of profit. As challenging as the business world can be, its mission is not nearly as encompassing as that of a responsible government. Its employees are spared some of the onerous rules that make certain actions more difficult for government officials. In some industries, a company can often survive by keeping its eye on the next quarterly or annual targets rather than planning for what it might face in the coming years and decades. It can also choose to go out of business. If a government chooses to do that, it's a really bad day for everybody.

So there you have it. Shaping the future of society, or at least enabling a better future, is the signature responsibility of government leaders and institutions, especially at the national level, because they're the ones who are supposed to be attending to long-term interests, not worrying about profits or quarterly financial reports. The only problem is that most of the time, they're not doing it. Passage of insightful, forward-looking actions that will span generations—and are adopted only after thorough consideration and appreciation of the long-term implications—is a rare occurrence that is usually preceded by a long, arduous process. Why is this true? Are the people involved lazy or short-sighted? Do they not care? Or are they simply overwhelmed with the short-term demands overflowing from their inbox?

There's more to this than the foibles of particular individuals or the heavy workloads of certain organizations. The underlying problem is the nature of the system we've established over many generations. It's loaded with incentives that generally drive us to short-term rather than long-term solutions. As stated in a November 2002 report to the president and Congress by the Commission on the Future of the U.S. Aerospace Industry (also known as the Aerospace Commission): "The federal government is dysfunctional when addressing 21st century issues from a long-term, national, and global perspective." The Aerospace Commission was focused on what's needed to sustain a strong U.S. aerospace industrial base, but its call to "reorient government organizational structures" rings true across the board. Put simply, the nation needs to create an environment that maximizes incentives—and minimizes disincentives—to think and act based on long-term interests.

As many have said before, we have a system that works away from problems rather than toward goals. But a different approach is required for tackling the long-term challenges facing humanity in timeframes measured in years and decades, not weeks or months. There's no such thing as instant gratification when you play the game at this level.

Public policy is often event-driven, which encourages policy-makers and the bureaucracy to be reactive rather than proactive. Some events are more predictable than others because they operate on a fixed schedule. The most obvious predictable policy-driver is elections. We know there will be a congressional election every two years and a presidential election every four years, which gives us the timing even though we can't always guess the outcome. But given what we know about the positions of the two major parties and the difficulty of making profound changes quickly, we can narrow the field of possible policy outcomes considerably.

Many other policy drivers leave us speculating as to their timing and impact, or sometimes are totally absent from our radar screens until they're upon us. These can result from significant changes—good and bad—in a variety of geopolitical, economic, and technological areas. In an age of weapons of mass destruction, global transportation networks, and dependence on interconnected electronic information systems, the technology factor has become more central than ever before in human history, and its tempo keeps increasing. Major new capabilities like the Internet can vastly expand opportunities at many levels of society, unfortunately including nefarious activities. Legacy systems like those for electricity or water distribution or space launch can fail, requiring quick decisions on resource allocations to either repair or replace them. Foreign entities can develop or adopt new technical capabilities, which can be good news or bad news depending on whether or not they're friendly.

Given adequate resources and a well thought-out set of options, policy-makers and the bureaucracy undoubtedly would prefer to be proactive more than they are now. Sometimes this works, and we manage to circumvent an immeasurable amount of pain as a result. Remember the millennium bug? If you don't, that's a good thing, because it means nothing happened, which was the whole idea behind an impressive, widespread effort across the public and private sectors. The last few years of the 20th century were spent correcting a software glitch that could have befuddled computers around the world as their internal clocks turned from 1999 to 2000. But as the critical time came and went, your electricity didn't go out, your bank account didn't disappear, and your Social Security account didn't suddenly show your age as a negative number. Was the whole thing a false alarm? Actually, it's more reasonable to assume that this non-event was caused by years of proactive planning and execution with adequate funding. Call it an unquantifiable bureaucratic triumph.

In addition to a preference for being proactive rather than reactive, there are other characteristics that help ensure a successful policy. You need to have buy-in from those who will provide resources to execute the policy (for example, the Congress and agency heads) and acceptance by the folks who will actually implement the policy (such as career bureaucrats). You also need a realistic timeframe to achieve the targeted policy objectives. Last but not least, you should have a very clear idea of what the policy objectives are. In general, programs are initiated to address new requirements or redress deficiencies in old programs. This could include any of the following: reducing costs, fixing schedule problems, improving the success rate, and inducing some beneficial change in the behavior of a target community.

Looking at the flip side, if it's bad policy decisions you're after, pay close attention because here are some of the surprisingly common practices that will take you in that direction. Note a common characteristic driving all of these: short-sightedness!

- **An unrealistic view of the achievable.** Sometimes this simply boils down to underestimating the required funding, personnel, and time. Another problem in this category is political infeasibility, which may be due to ideological disputes over the wisdom of a particular course of action, fear of undesirable foreign policy implications, strong resistance from advocates of competing programs, or political constraints or trade-offs imposed on an agency due to some unrelated activity. In areas like defense and space, another common problem is wishful thinking regarding how fast technologies will mature. Too often, project planners are overconfident that technologies in early stages of development will be ready for operational use by the time a new system is scheduled for deployment. Very often, they are disappointed and end up spending an inordinate amount of their budget and schedule trying to bring key technologies to the point of full readiness, resulting in cost overruns, schedule stretch-outs, and reductions in performance expectations.

- **The solution doesn't fit the problem**. This is a pitfall that can result in programs that are either too big or too small. Zealous planners, seeking to advance to the next-generation capability, may propose an overly ambitious solution to a problem, sometimes even before the problem has been clearly defined. For example, proponents may call for a whole new satellite system (or transportation system, or weapon system, etc.) when all that's needed is an upgrade of a subsystem or component. On the other hand, overly cautious planners may favor responses that are insufficient to the task if they are worried about the resource requirements of a big project, have a vested interest in preserving an old project, or want to make sure that any new action is small enough to be easily reversible. They fail to recognize that sometimes you really do need to move to that next-generation system.

- **Haste and poor planning**. Trying to lock in a project too quickly or shortchange the planning cycle contributes to the problems noted in the previous two paragraphs. Decisions are sometimes made based on inaccurate or inadequate information, usually due to insufficient consultation with subject-matter experts and the program's ultimate user community. Consideration of input from all key interest groups requires time and expense up front, but skipping this step adds time and expense—sometimes a considerable amount—later in the development cycle. Rushing to put something in place before an arbitrary deadline, or to undercut the efforts of a competing organization, can sabotage program success by failing to adequately explore likely scenarios and plan for contingencies, or by failing to identify conflicts with existing policy or law.

- **The desire to assert authority or to make a political statement**. This type of behavior almost guarantees bad policy when applied to decisions having long-range implications. Even if a policy-maker succeeds in achieving personal or political gain, it will be short-lived. If the decision ultimately serves the national interest, it will be by accident rather than by design.

Short-sightedness in decision-making has a cumulative effect that can put the nation in a very bad place, undermining important programs and the institutions that are needed to carry them out. Many would argue that we've already done this to such a degree that it will take many years of struggle to recover, if we ever can. Why would we let this happen? Because incentives throughout the system drive us toward short-term thinking. Borrowing a slogan from a government document I saw several years ago: we need to "incentivize robustness."

(I don't know who decided that the noun "incentive" should morph into the verb "incentivize." It could have been someone in the federal government who liked to speak in acronyms and abbreviations. Some of the sentences uttered by such people contain no real words other than verbs. Therefore, it's understandable that they'd be incentivized to create more verbs for the sake of variety.)

As we analyze the forces that drive the government toward short-sightedness, they seem to have their roots in three components of the political environment: the election cycle, the budget cycle, and the news cycle. Certainly, there's nothing revolutionary about recognizing the effects of these cycles on politics and policy. But we'll focus on how these cycles distract the Congress, the president, and the bureaucracy from their responsibility to shape the future by shortening their perceptions of the temporal horizon.

Incentivizing Congress

The nation's founders considered the legislature to be the "first branch" of government, so they addressed it in Article I of the Constitution. But their wisdom in creating this bicameral body of elected representatives didn't include immunity from misplaced incentives that can accumulate over time.

Political scientists have identified three primary goals motivating congressional behavior: getting re-elected, enacting good policies, and winning the respect of peers. The fact that re-election is Goal No. 1 should not be a surprise, but it's more than just a desire to keep the job and the prestige. Freshmen members of Congress quickly realize that their effectiveness depends on being around long enough to learn their job, build their networks, and hopefully work their way into leadership positions. That's not something you can accomplish in one term, especially in the House, where the term length is just two years. Once you get yourself established, then you're in a position to pursue the other two goals: enacting policies and winning respect. If you're reasonably successful at doing this, you're motivated to stay longer and hopefully achieve more. In other words, the longer you stay, and the more successful you are as a policy-maker, the more you have to gain by getting re-elected. Sounds logical—but how much of your time and effort do you have to devote to retaining your job?

In the 2008 election campaign, the average cost of running for a seat in the House of Representatives was $1.1 million. Amortizing this investment over the two-year term, that's $550,000 each year for a job that pays just $174,000 (for rank-and-file members as of 2009; House leaders get $193,400). Unless you're extremely wealthy and you like losing money, that means you've got a lot of fundraising to do, and not much time to do it. For the average House incumbent running for re-election, the campaign fund has to grow at a rate of about $1,500 per day throughout the term. If re-election is indeed the top goal, then obviously the fundraising chore has to be a top priority.

Maybe it would be better to run for the Senate. It's more prestigious, and the term lasts six years. But Senate campaigns in 2008 cost an average of $6.5 million, which works out to $2968 per day—almost twice the daily fundraising requirement of the House, despite a term that's three times as long. And the Senate's pay scale is the same as the House.

Congress-members don't have the luxury of ignoring their campaign war chest. The cost of campaigning for national office has been rising steadily. Campaigners are well aware that in recent congressional elections, the candidate who has outspent his/her opponent has been the winner in an overwhelming majority of races, with success rates ranging from a low of 73 percent (Senate, 2006) to as much as 98 percent (House, 2004).

In today's environment, big money is essential for connecting with voters and getting a message out, but that's not the only thing needed to ensure re-election. Constituent service throughout a member's term is also important. However, the majority of personal contacts received from constituents do not convey a message like, "Please support a long-term strategy for [research or infrastructure development or environmental protection] that will benefit the nation and the world for many generations to come." More often, constituents make requests for help on an immediate problem, or express support or opposition to a current issue that affects an individual, their district, or their state. Once again, the member is compelled to focus on the short term.

Considerable attention has been given to the campaign finance problem in recent years, although the focus usually is on moderating the influence of big money interests rather than helping elected officials eliminate incentives for short-term thinking. We've observed for many years how hard it is to modify the system because donors, candidates, and political parties always rebel against changes that they believe will put them at a disadvantage. Nonetheless, if campaign finance reform eventually succeeds in mitigating the fundraising pressures on candidates through spending caps, public funding of campaigns, or some other method, that will help take some pressure off members so they can spend more time focusing on timeframes beyond the next election.

As hard as it is to enact campaign finance reform and make it stick, it's even harder to change the Constitution, as demonstrated by the fact that it has gained only 27 amendments in well over 200 years, the first 10 of which were passed as the Bill of Rights at the very beginning. A successful amendment requires a two-thirds majority in both houses of Congress and approval by three-quarters of the states. That's what would be required to lengthen the absurdly short terms of House members. The drafters of the Constitution picked two-year terms because they envisioned the House of Representatives as a body consisting of white male property owners who would serve one or two terms of public service and then return to their farms and businesses. Reality diverged from that model so long ago that it's surprising we haven't increased House terms to at least four years by now. That way, we could give members some breathing room so they can concentrate more on governing and planning and less on fundraising and campaigning.

The president, the members of the Senate, and 48 of the nation's governors have terms lasting at least twice as long as those of the House, so why should two-year terms continue to be forced upon such an important legislative body? Two-year election cycles waste money providing opportunities for choice at a frequency that obviously is not needed, given that incumbents running for re-election win their races over 90 percent of the time. Although fundraising is a big part of that electoral success record, it's also an indication that constituents aren't all that eager to change their representative in most cases. As pollsters have consistently noted, the Congress as a whole tends to get a low approval rating (around 30 percent is not uncommon), but poll respondents give very favorable ratings to the representative of their own district. Apparently, the low rate of turnover in the House is largely due to the fact that the voters who say "throw the bums out" are almost always referring to bums from districts other than their own.

Undoubtedly, a lot of time and effort is devoted to the re-election goal. But the work of legislating is the real reason that the member was sent to Washington, and therein lies the opportunity to shape the future. The tools to do so are many and powerful, and the biggest one is the power of the purse. A new member of Congress will quickly perceive the immense potential of budgeting for channeling benefits back to the home district (another contribution to the re-election goal), for advancing U.S. security and economic interests, for relieving the hardship of those in need, and in general for making the future a better place than the past. As one learns more about the budget process, however, two things become evident: it takes up an inordinate amount of Congress's time and effort, and it has evolved into a process that makes forward-looking policy-making far more difficult than the founding fathers could have imagined.

In the 1980s, political scientist Aaron Wildavsky pointed out that budgeting had developed to the point that instead of being merely a subset of governing, budgeting had *become* governing. He found that by the early 1980s, more than half of the floor votes in the House and nearly three-fourths in the Senate dealt with taxation, authorization, appropriations, and debt ceilings. And that doesn't count the time spent in numerous committees.

All that time spent on the budget is worthwhile if it provides the resources and guidance needed to carry out a well-developed strategy for serving the nation's current interests while building a better future. Unfortunately, that seems like an impossible dream when the majority of the federal budget—62 percent in 2008—is devoted to so-called mandatory programs. These are programs with outlays that are determined by law or fiscal obligations, such as Social Security, Medicare/Medicaid, other pensions and entitlements, and interest on the national debt. Since the nation's revenues consistently fall short of outlays, resulting in deficit spending, and the mandatory programs are essentially on autopilot, the discretionary programs—the other 38 percent—must live by rules intended to enforce fiscal discipline.

Obviously, the rules haven't succeeded in avoiding deficit spending, except for a brief period from 1998 to 2000. (Even during this period, only the 2000 budget produced a surplus without the help of offsets from the Social Security surplus, which is generally considered an accounting trick.) For purposes of our discussion on shaping the future, a key characteristic of this environment is that long-range projects are harder to initiate, especially if they have substantial start-up costs, because they break through budget caps now and in the future, which makes deficit targets harder to achieve and disrupts resource planning for existing programs. The result is that members, aligned with agency advocates, rally around existing programs to the detriment of new ones, despite the potential for innovation that new programs may have. Alternatively, new programs that manage to get approved will be introduced with inadequate start-up funding and unrealistically low estimates for out-year spending, causing higher costs and schedule delays down the road. This behavior is occurring at a very unfortunate time: an era that is demanding exceptional levels of innovation in the search for solutions to national and global problems.

The U.S. has experienced deficit spending in almost every year since 1970, so this is not a new problem. Since that time, the Congress has attempted to discipline itself on several occasions, such as the extensive congressional reform actions of 1974, the two Gramm-Rudman-Hollings deficit reduction acts of the 1980s, and the Budget Enforcement Act of 1990. But deficit spending persists, as does the tardiness of annual budget bills. Meanwhile, the budget process becomes more complicated, and in the words of Aaron Wildavsky, "guarding against its own worst tendencies in advance, Congress anticipates its collective unwisdom by taking away its discretion." Indeed, for a long time some commentators have charged that Congress has abdicated its power of the purse through its increasingly byzantine procedures and its unwillingness to make hard choices about reforming mandatory spending. Frustration with this situation has been voiced inside as well as outside the Congress. To conduct an extensive examination of its rules and procedures in 1993, the Congress created the Joint Committee on the Organization of Congress, which assessed reform proposals in a number of areas, including the budget. The Senate version of the Joint Committee report noted that:

> . . . although a great deal of time is spent on the budget, little time is spent in long-term planning, overseeing

programs, and finding waste and abuse. In short, the Congress spends too much time on budgetary issues that do not matter and not enough time on those that do.

Both the House and Senate versions of the Joint Committee report recommended a shift to biennial budgeting. The proposed two-year cycle would schedule budget resolutions and appropriations bills for the first session of each Congress and authorization bills for the second session. Reasons cited for taking this step included relief from unmanageable committee workloads, freeing up of committee resources for oversight duties, elimination of redundant budget-related votes, encouragement of greater stability and longer time horizons in policy planning for both committees and agencies, and relief for agency officials from the never-ending cycle of annual budget preparations.

There are other ways of implementing a two-year cycle besides the one suggested by the Joint Committee. For example, the Congress could retain its current procedures, but only address half the budget each year. Although options abound, none have been adopted because members believe they would lose control of both the spending and the programmatic behavior of executive agencies, and would be confronted with greater numbers of supplemental budget requests in the off-years. However, the reality could turn out to be just the opposite. Congress could have *more* control because committees would actually have time to do program oversight. That oversight, combined with additional time to do proper long-term planning, could *reduce* the number of supplemental requests due to agencies' increased efficiencies and effectiveness as a result of the longer cycles.

Congress-watchers have identified two types of committee oversight. One is "police patrol," in which committees cruise around the agencies and programs in their jurisdiction, checking to make sure everything is okay, and taking action if something is amiss. The other is "fire alarm," in which committees react to emergencies in their jurisdiction, sometimes helped by legislatively mandated "alarm bells" that alert them of problems (for example, a program that is more than 15 percent over budget). As you can probably guess, the fire alarm method—the reactive approach rather than the proactive one—is more popular. Under most circumstances, congressional committees don't have the time or resources to conduct police patrol oversight, in part because budget actions are taking too much of their time.

The reason Congress receives supplemental budget requests from agencies is that programmatic needs are often unpredictable. This is obviously the case, for example, when responding to unexpected natural disasters that require rescue and recovery efforts, but is also true of government functions that involve research and development (R&D), large-scale infrastructure projects, environmental protection and cleanup, and unique procurements such as military systems. For these types of activities, yearly cycles are a very poor fit. Government programs follow annual budget cycles only because we force them to. The length of the budget cycle is not specified in the Constitution. In fact, Article I, Section 9 gives only this general guidance: "No Money shall be drawn from the Treasury, but in Consequence of Appropriations made by Law; and a regular Statement and Account of the Receipts and Expenditures of all public Money shall be published from time to time."

It has been customary for the president's annual budget request to include a projection five years into the future to help the Congress understand the administration's intentions for the evolution of programs and to provide some indication of the resources required beyond the immediate budget year under consideration. The first budget request of President Barack Obama in early 2009 included a 10-year projection for the first time, which would seem to be a positive step. Unfortunately, the Congress, under the leadership of his own partisans, chose to ignore this change, and stayed with its traditional five-year perspective.

Incentives for short-term thinking, dominated by the election cycle and the budget cycle, create both subtle and overt hindrances to the ability of the Congress to focus on the long term. Whether the subject is research, infrastructure development, regulation, or other issues that reach beyond the two-to-five-year comfort zone, the legislature frequently resorts to two short-term solutions: delegate the tough decision-making to an executive agency, and/or postpone the issue to a future Congress. Within the realm of space policy, examples are easy to find.

- In 1984, as prospects for commercial space efforts emerged in industries such as launch services and satellite remote sensing, the Congress took three key actions. One was to amend NASA's charter, requiring the agency to "seek and encourage, to the maximum extent possible, the fullest commercial use of space." It was left to NASA, which up to that time had no mandate to facilitate space commerce, to define what that meant and figure out how to implement it—without being given any specific funding to do so. The second and third actions taken that year were the two laws passed to permit commercialization of space launch services and remote sensing. To make a long story short, both laws failed to give executive agencies the tools they needed to enable an adequate business environment. No significant commercial activity emerged in launch services until the launch legislation was substantially amended in 1988, or in commercial Earth observation until the remote sensing law was repealed and replaced in 1992.

- The struggle for suitable legislation for the commercial space launch industry continues today. The 1988 amendments to the launch act previously mentioned included U.S. government provision of third-party liability indemnification for catastrophic launch accidents. Since that time, there have been no such accidents, so there has been no cost to the government. Nonetheless, the Congress refuses to make that provision permanent, preferring to bring it up for renewal every few years, which adds uncertainty to business plans. Foreign competitors to U.S. launch firms already have similar indemnification through permanent legislation in their host countries.
- NASA and other proponents of human exploration often point to the NASA authorization bills passed in 2005 (covering fiscal years 2006-08) and 2008 (covering fiscal year 2009) as proof of bipartisan endorsement of lunar and Mars missions. The 2005 bill calls for a return to the Moon by 2020, followed by a "sustained human presence on the Moon . . . to promote exploration, science, commerce, and United States preeminence in space, and as a stepping-stone to future exploration of Mars and other destinations . . . on a timetable that is technically and fiscally possible." The bill requires NASA to devise a solar system exploration plan "that is not critically dependent on the achievement of milestones by fixed dates." Other than the return to the Moon, this commitment is vague and easily reversed. Indeed, the language in the 2008 bill begins to back off on the pledge, leaving the real decisions about long-term exploration to future Congresses: "Congress hereby affirms . . . the *broad goals* of the space exploration policy of the United States, including the *eventual* return to and exploration of the Moon and other

destinations . . ." (emphasis added) "The timetable of the lunar phase of the long-term international exploration initiative shall be determined by the availability of funding." "NASA shall make no plans that would require a lunar outpost to be occupied to maintain its viability. Any such outpost shall be operable as a human-tended facility capable of remote or autonomous operation for extended periods."

To some extent, the lack of specificity in legislation on the implementation of long-term policy is understandable. The Congress does not have expertise and resources comparable to the executive branch for highly technical, forward-looking planning. Aside from the staffers of members and committees, its research support includes the Government Accountability Office, which audits the performance of agency programs; the Congressional Budget Office, which looks forward but only with regard to federal spending and revenues; and the Congressional Research Service, part of the Library of Congress, which does non-partisan summaries and analyses of policy issues at the request of the Congress. From 1972 to 1995, there was also the Office of Technology Assessment (OTA), which produced hundreds of studies, often in considerable depth, on science and technology issues. The OTA was eliminated in a symbolic budget-cutting gesture after the Republicans took control of the Congress in 1995 (although the OTA's annual funding of approximately $20 million was simply shifted elsewhere in the congressional operations budget). The demise of OTA is another example of shortsightedness driven by electoral and budget pressures.

The task of adding technical specificity to legislative mandates falls to executive agencies. Similarly, the task of overcoming the vagueness in congressional statements of purpose associated with forward-looking programs should fall to the presidency. But can the president be counted on the pick up the mantle?

The president as futurist

The circumstances described so far indicate that members of Congress have strong incentives to think short-term—focusing on a one-year, two-year, or at most five-year time horizon—despite the fact that so many of them hope for a much longer tenure in the legislature. That leaves it up to the president (or more appropriately, the institution of the presidency with all its advisory and support mechanisms) to pave the way to the future. But the president doesn't expect to be around for more than two four-year terms, a limit that's been imposed since the 1951 ratification of the 22nd amendment to the Constitution. Does the two-term limit prevent the office-holder from thinking about and planning for what happens beyond his/her watch?

Presidential scholar Paul Light has suggested guidelines on the motivations of a president that are very similar to those mentioned earlier for congress-members. He identified the president's three major goals as re-election, good policy, and historical achievement. Note that the first two are the same as for the folks on Capitol Hill, but the third one differs. Rather than seeking the respect of their peers, presidents aim for a bigger prize: the respect of history. This would seem to indicate a focus on lasting accomplishments for the benefit of the nation and the world. Take heart, futurists—we may be onto something here! But before jumping to any conclusions, let's consider how long-term planning (in a timeframe of decades, not years) is shaped by all three of the president's goals.

Any president who expects to run for a second term will spend the first term looking ahead to the looming re-election campaign. The president has short-term concerns similar to those already discussed for congress-members: growing campaign costs (exceeding $1 billion for the first time in U.S. history for the two major candidates combined in 2008) and the need to provide visible benefits to constituents, especially in states with lots of electoral votes. An additional concern is that presidential campaigns are getting longer, with the initial posturing beginning more than two years prior to the election, which is part of the reason why the campaigns cost more. Barack Obama set a new record in the 2008 election cycle by spending $730 million, which translates to an average of just under $500,000 *per day* amortized over a four-year term. Compare that to the president's salary of $400,000 *per year*.

The president's other two goals, good policy and historical achievement, are closely linked, but a president's best intentions in these areas are often confounded by current events. For example, when Richard Nixon entered office, he wanted to do foreign policy and let the country run itself. Instead, he found numerous domestic issues on his plate, such as the economy, the environment, energy policy, and civil unrest over the Vietnam War. In another example, Bill Clinton told voters during his 1992 campaign that as president he would "focus like a laser beam" on the economy. Conflict in Kosovo, the rise of international terrorism, and relations with Russia and other former Soviet states were among the distractions that altered his focus.

As noted earlier, government investments in long-term projects such as space research and development may or may not yield visible benefits, and even if they do, the benefits are likely to be far in the future, widely dispersed, disassociated from their origin, and impossible to measure. For a first-term president, championing such projects does nothing to support the re-election goal, and may even be a hindrance if large up-front expenditures are visible and benefits are not. This problem goes away for a second-term president, but that doesn't necessarily mean smooth sailing for big long-term visions, which still need to demonstrate their contributions to the president's policy and historical achievement goals.

Big visions with big price-tags and unsure benefits that are far in the future, like the exploration and development of space, fall into a category that Paul Light has labeled the "constituentless" issue. Interest may be limited to a narrow community, and even if widespread public interest develops, it's likely to be shallow and fleeting. This doesn't encourage the formation of major, sustainable support coalitions. Therefore, a constituentless issue is less likely to attract strong, sustained support from a president because it doesn't clearly support the three primary goals, and may promote conflict among them. The president is motivated to avoid risky ventures in the annual budget requests because rejected or canceled projects don't reflect positively on a president's record. Imagine how differently we'd view John F. Kennedy if the Apollo program had been terminated during or shortly after his time in office.

Speaking of Kennedy, his initiation and active promotion of Project Apollo during what many believe to have been the most dangerous period of the Cold War seems to run counter to the argument made so far. Indeed, Kennedy is the exception that proves the rule, an example of presidential science and technology leadership occurring during anomalous circumstances that opened a window of opportunity. As a result of this example, there has been a persistent belief in the space community, especially among citizen advocacy groups, that the most important factor in formulating and implementing a productive, forward-looking space program is the enthusiastic support of the president. The validity of this Apollo-era leadership model in today's environment is highly questionable, as we've seen.

More valid today is the persuasive presidency model advocated by many political scientists. I'll briefly draw from one of them, James

MacGregor Burns, who wrote a book on presidential leadership (which he cleverly titled *Leadership*). According to Burns, when a president wishes to substitute new goals for old ones, the new goals must have equal legitimacy as well as political validity. This is difficult enough in established organizations with long-accepted goals, but it's even more difficult when the new goals entail significant innovation. Lacking an established support coalition, the president must act as the persuader, not the decider. To use Burns's terms, civil space projects since the end of Apollo, as well as much of the nation's other scientific and technological efforts during that time, have been experiencing *transactional* leadership (satisfying individual or group needs along the way—an incremental approach) but not *transforming* leadership (aiming at higher collective goals).

It may seem odd that science and technology could be considered constituentless issues. After all, we routinely hear commentators, from the citizen on the street to the talking head on television, assert the importance of science and technology in our daily lives and our future. But all too often, there's little substance beneath that veneer of attention. Polls always show that the citizens' biggest concerns, not surprisingly, are the immediate ones: the economy, overseas conflict, health care, crime, and entitlements like Social Security. (In Chapter 8, we'll consider how less-immediate concerns can move higher on the public's agenda.)

Innovative changes and large scientific and infrastructure investments have become increasingly difficult to justify if they appear isolated, in substance or in time, from current societal needs. So the presidency and the Congress—the two institutions responsible for looking after the nation's long-term future—are overwhelmed by incentives to think and act for the near-term. They work away from problems rather than toward goals. And current events keep them well supplied with problems.

Keeping up with the news

Much has been written about the changing face of journalism. As television became a fixture in American homes starting in the 1950s, the video-enhanced daily news reports, political events, and campaign ads drew an ever-larger audience and more media competition into the news cycle. The global satellite links established in the 1960s and 70s gave birth to real-time reporting from around the world. By the 1980s, cable TV was becoming popular in the U.S., making 24-hour news channels a prominent element of American viewing habits. The 1990s, of course, added the Internet to the mix, and this medium continues to evolve. Today, political candidates at all levels of government would be wise not to underestimate the importance of having a good campaign website, and should not ignore the possibility that anything they do or say in public (and sometimes in private) could end up being posted as a video clip for all to see.

Research has shown that candidates' sound bites on the news media during election campaigns in the U.S. have gotten shorter over the years, dropping from an average of 42 seconds during the 1968 campaign to barely eight seconds in 2004. That means the messages politicians convey through the broadcast media have to be crisp, whether they're campaign pitches or statements on issues in the normal course of business. Other media that give the policy-maker some control, such as a personal website, provide an opportunity for more depth, but not much, since statements are typically no more than 300 to 400 words so readers won't lose interest. Continuous updates are important as well, because yesterday's posts are old news. In an environment where information is dispensed quickly in small packets, how is a policy-maker, a futurist, an academic, a government agency, a think tank, or anyone else going to successfully propose, promote, and develop a multifaceted vision for the long-term future? As individuals and as a society, we're still at the explore-and-adapt stage with new media capabilities. Though younger people have grown up with the Internet and may take it for granted, it's still a relatively new medium and we're still at an early stage of the learning process. Rather than sharpening our view of the distant horizon, new capabilities that continue to evolve rapidly are making that view more fuzzy, at least for the time being.

Newspapers used to be the best, most accessible way for most people to follow issues on a routine basis and in a reasonable amount of depth. Unlike most TV and radio coverage, the stories could be long, and could appear in a series of articles over days, weeks, or longer. A classic example is the reporting on the Watergate scandal by Bob Woodward and Carl Bernstein. In case you missed it—which probably means you weren't born yet—the reporting included numerous articles in the *Washington Post* as well as two books between 1972 and 1976. That episode was largely responsible for adding the word "investigative" in front of "journalism" in our lexicon.

For many decades, the daily paper survived competition from radio, and then television, even as these electronic media grew enormously in the number and variety of channels offered. The Internet presents a different kind of challenge to newspapers, undermining their print editions if not their traditional concept of news coverage. Obviously, the Internet's content and reach grew much faster than its predecessors in electronic media. Instead of going from a few channels to hundreds of channels in a few decades, the Internet grew to untold millions of informational websites and spread around the world in a few years. The websites can be interactive, can be updated on a continuous basis, and can present live or recorded videos. They can be accessed from a desktop at school or the workplace, from a screen at home, or from an array of mobile devices. Accessing websites doesn't require physically seeking out a news vendor and buying a bundle of paper, or arranging to have that bundle of paper routinely delivered to your house, where it often gets lost in the bushes or soaked in the rain. So it looks like print newspapers are succumbing to superior technology and will go the way of typewriters and 8-track tapes. (If you've never heard of typewriters or 8-track tapes, look them up on the Internet.)

So what's the problem? We still get the news, we get it from more sources, we get it updated more frequently, and we get it via electrons so that we don't have to kill trees and create wastepaper to stay informed. At first glance, it may seem that traditional newspapers are poised to be big winners in the move to the Internet. They have the organization, the experience, and the name recognition. Their web presence, in many cases, is attracting significantly more readers than their print editions. But news organizations survive on advertising revenue. Even though ad income for print editions has been down in recent years, it still exceeds online advertising by a substantial margin. This may seem counterintuitive, since advertisers typically direct their money to wherever the biggest circulation can be found. However, print newspapers have reasonably good information on the demographics of their readers, helping advertisers target particular audiences. Online news typically has little or no demographic information on its readers, which discourages potential advertisers. Also, website visitors spend very little time in one place before moving on. The *Washington Post* reported that in March 2009, readers on its website spent an average of 16 minutes, compared to 30 minutes reading the daily paper and 60 minutes for the Sunday edition. A large percentage of online readers—more than half in the *Post*'s March 2009 assessment—entered the site using a link from another site or search engine rather than through the site's home page. This means that a lot of users, possibly a majority, come to a news site to quickly find specific information, not to use the site as their daily window on the world.

The problem this creates is that the competitive, fast-moving, and rapidly evolving environment compels news outlets large and small to do everything they can to grab and hold our attention. They have an incentive to sacrifice accuracy and depth for the sake of speed and brevity. (Getting the scoop before the competition is now a matter of seconds, not hours or days.) There's a temptation to sensationalize or to focus on the flashy stories expected to generate plenty of activity on search engines, and downplay the mundane stories even when they're important. In the "drink from a fire hose" environment of today's media, how can the average person sort out the good from the bad, or the essential from the useless, and do it without expending too much time? This is especially challenging for young people whose filtering skills are still developing, but it can be tough for anyone. Sturgeon's Law (named for science fiction writer Theodore Sturgeon) tells us that 90 percent of everything is crap. If that's true of the Internet, then we've all got a lot of crap to wade through.

Speaking of which, here's an example of how the news environment is morphing faster than we can keep up. A few years ago, no one had ever heard of blogs. Now everyone has one, or wants one, or participates in one or more. We're even told (by the media, of course) that blogs have become politically influential. This is both encouraging and scary. On the encouraging side, the "democracy" of the Internet gives everyone a voice in all the issues of the day. Whether you're trying to sell your perspective or comment on someone else's, you're welcome to participate. Your ideas may influence the thinking of others, or maybe just the act of writing your thoughts will help you refine them, in which case you've influenced yourself. If this is the essence of blogging, then it's a step toward better democracy through more active participation.

On the other hand, Sturgeon's Law comes into play. That's the scary part about blogs becoming politically influential. If Sturgeon is right, his 90 percent rule puts an unsavory twist on the concept of majority rule. An unscientific sample based on my own online experience tells me that in blog discussions, and in reader comments posted to news stories, an alarmingly large percentage of the remarks are negative and/or shortsighted. Also, in the science and technology policy topics that I tend to follow, it's disturbing to see how ill-informed many respondents are about all aspects of the subject matter. Presumably, they don't intend for their postings to demonstrate their ignorance to the world, so apparently they think they're making an informed, valuable contribution to the argument. That doesn't bode well for producing constructive results from broad participation in long-term planning and decision-making.

Some might offer this rejoinder: "It doesn't matter. The Internet is a self-correcting system. If someone posts something inaccurate, someone else more knowledgeable will post a correction." Such an assertion does not indicate an appreciation for the extent of the problem. Bad information can spread very quickly once it gets linked to other web pages, circulated by email, or indexed in a search engine. The corrections won't necessarily follow the same path, so the erroneous information becomes conventional wisdom, and damage control has little or no effect.

Internet users see this all the time. People post inaccurate information, deliberate lies, and personal insults that they would never attempt in face-to-face communication, or at least would not be able to do without being challenged by their peers. A disgruntled employee posts malicious untruths about a former employer. A political partisan falsely attributes words or deeds to a candidate he dislikes to make the candidate appear stupid or malevolent. This behavior thrives during election seasons. For example, Vice President Al Gore was a favorite target during the 2000 presidential campaign. Most people still think he said "I invented the Internet" when in fact he never made such a claim. Few have bothered to explore multiple sources—so easy to do on the Internet—to learn that he does deserve credit for championing high-speed computing and networking in the Congress starting in the late 1970s, and culminating in the High Performance Computing and Communication Act of 1991, which paved the way for the Internet to extend its capabilities and expand its user community.

Also during the 2000 campaign, an email made the rounds that contained a list of dumb quotes, all attributed to Gore. I recognized some of them and knew that they came from a different source. Here's my favorite:

> Mars is essentially in the same orbit. Mars is somewhat the same distance from the Sun, which is very important. We have seen pictures where there are canals, we believe, and water. If there is water, that means there is oxygen. If oxygen, that means we can breathe.

The actual speaker of this and other quotes attributed to Gore was Vice President Dan Quayle, who chaired the National Space Council during the first Bush administration. Quayle was responding to a question from a CNN interviewer in August 1989 on why the U.S. should send people to Mars. I remember hearing about this at the time it occurred, but you don't have to take my word for it. Check multiple sources—a good practice under any circumstances, although sometimes it can take some effort to find multiple *reliable* sources.

Fact-checking with multiple sources doesn't always guarantee you'll find the best sources. Most subject-matter experts are too busy to spend time on the hopeless and thankless task of correcting errors on the Internet related to their field of expertise. Interaction among experts often takes place via email or password-protected discussion groups rather than open forums. Knowledgeable people in government and industry have restrictions on what they can say in public, and on their level of participation in online activities. They keep a low profile so they won't inadvertently reveal sensitive information, and so no one will get the false impression that they're speaking for their agency or company. But if all these smart folks who really know the inside scoop refrain from joining the discussion, who's doing all the blogging? Sometimes it's really sharp people who know their stuff and have more freedom to speak, such as academics, think-tank researchers, and retired experts. But many times, it's enthusiasts or detractors who lack the depth of knowledge and the credentials to be reliable authorities. It's not always easy to tell the difference, especially when people misrepresent themselves. You can't just assume that the names you see most frequently on the web are the ones you can trust, since the most prolific writers may turn out to be the ones with the most freedom to speak and the most free time, not the most knowledge.

Some readers may interpret this discussion of the electronic media as Internet-bashing by some old-timer who gets excited about the feel of newsprint and owns stock in failing newspapers. Actually, I've long been an advocate going paperless to the greatest extent possible, and the Internet is obviously a big part of that. In fact, it's quite possibly the most important information tool since the printing press, which first made news and books available to the masses. Every day, I find new ways the Internet keeps me informed and connected. I also recognize that it's far from perfect.

There may be those who believe the Internet's dangers exceed its benefits, providing fertile ground for criminals, terrorists, and pornographers. Or they may see it as a major contributor to the "dumbing down" of society, catering far more to consumerism and entertainment than to intellectual pursuits. I take a different view. I believe the Internet is still in its infancy, and so is the way we use it and respond to it. The ability to reconfigure the way news is presented, while simultaneously speeding up the cycle, affects the behavior of users and of other types of media. For a while, the tendency of web developers will be to focus on immediacy and on bite-size, easily digestible reporting. That adds to incentives for short-term thinking. But eventually this will change, as the Internet goes from being mostly a source of quick answers to being a source for most everything. There's no way of knowing how long it will take to get from here to there, so I like to think of this as a time of literacy in transition. In the meantime, even if Sturgeon's Law holds true, I can ignore the 90 percent that doesn't interest me, just as I can switch channels on my TV or radio. The remaining 10 percent is my treasure, my intellectual enrichment. I'm sure I'm not the only one who filters the daily onslaught of information in this way. The optimist in me resists the notion that the net effect of the Internet is negative, although my cautionary side says we need to remain vigilant to keep from succumbing to the dumbing-down effect.

An important tool for measuring the effects of current events and the media—and another influential component of the news cycle—is public opinion polling. Polls attempt to take a snapshot of what people are thinking at a particular point in time. Current events alter poll results over time, sometimes quickly and dramatically, so a single poll never tells the whole story on what the population wants or believes. If the same questions are asked in numerous polls over a period of months or years, a better representation of the population's mindset takes shape.

When politicians say "I don't pay attention to polls," they're making a statement about the strength of their own convictions and signaling that they will not be swayed by fads or political pandering. But they actually do pay attention (or at least, their staff do) to cumulative results that can indicate important trends. They would be unwise to completely ignore polls for two reasons: polls are tools that support their re-election goal, and polls provide potentially useful (and free!) information on evolving policy issues.

Does this help long-term strategic planning for issues like the exploration and development of space? Unfortunately, issues like space policy are "low salience," to use a political science term. Civilian space projects have never been significant campaign issues, and the political parties have not engaged in organized efforts to take sides, at least not in ways that would be visible to voters. As a result, the public—even the attentive public outside of active members of the space community—has little incentive to get involved in the political process for space issues. Polls have confirmed the notion that the public typically has depended on elites to set the government's space agenda.

Members of Congress have been found to be most responsive to constituents, playing the role of faithful delegates, when issues are salient, signals from the constituency are clear, and consequences are traceable to the individual member's actions. But these conditions typically are not true in the case of space policy matters. In the absence of clear constituency positions, members who choose to consult public opinion polls on space issues find very limited guidance. The polls are national in scope, not limited to the member's state or district, and they hold few clues as to how space funding should be allocated. For example, respondents tend to be evenly divided on the relative importance of human vs. robotic missions, but in general they prefer scientific return over space spectaculars like a piloted mission to Mars. These vague preferences have remained essentially the same in polls since the 1960s, except for brief peaks in support at the times of the first Moon landing and the space shuttle accidents. The demographics of the respondents taking each position have remained consistent as well: space supporters tend to be white, college-educated males with incomes above the national household median; indifferent or unsupportive attitudes are most prevalent among women, minorities, people with less than a college education, and those with lower-than-median incomes.

Three decades of surveys by science and technology pollster Jon D. Miller have cast doubt on the reliability of the public's assessment of the value of the space program. Miller has found that the interested and attentive public displays disappointing results on questions of science and space literacy, and the performance of the inattentive public on these questions is significantly worse. In short, even those who like the space program and value its scientific advances would have a difficult time explaining what they've gained from it and why they feel as they do. People in the space community often say, with some justification, that public support for the space program is a mile wide but only an inch deep. Miller, like many other analysts, advocates improvements at all levels of public education on science and space, lest the public continue to shun involvement in decision-making on space issues.

These findings regarding the general level of public knowledge and interest suggest that most elected representatives are not acting as faithful delegates (doing the direct bidding of their constituents) when it comes to space policy and other technical matters. Rather, they are behaving either as trustees (left to their own judgment by a trusting electorate) or party conformists (voting with their partisan bloc). It's debatable as to whether this is good news or bad news for long-term strategic policy. One could argue that political elites, left to their own devices, will do a better job of planning and decision-making without the involvement of a populace that doesn't understand and doesn't care. On the other hand, given the distractions and short-term incentives identified earlier, one should not be overly confident of a high percentage of positive outcomes, which may result more by accident than by design.

* * *

This chapter has identified processes that inhibit long-term thinking among the nation's decision-makers: election cycles, especially the two-year terms in the House of Representatives; the rising cost of election campaigns; the need to attend to the short-term needs of constituents; the annual federal budget cycle, characterized by chronic deficit spending; the acceleration of the news cycle and the continuing evolution of the media; and the mixed (or absent) signals from constituents regarding technical issues like space policy. These things are very difficult to fix, requiring actions such as amending the Constitution, altering congressional procedures, enacting and enforcing campaign finance reform, and in the case of the budget deficit, making hard choices and sticking to them despite the political consequences. Evolution of the news media demands adaptation rather than fixing. All of these actions should be pursued even though they are uphill battles. But let's be realistic. The clear need for action has been around for decades. In some cases, reforms have been attempted but they have failed. Changing the process will take considerable time, and the issues won't wait. Progress needs to be made even while we continue to push for updates to entrenched rules and behaviors.

Rather than dwelling on the procedural obstacles in the system, some critics of Congress have observed, particularly in relation to the budget, that "the process isn't the problem, the problem is the problem." By this they mean that existing mechanisms, though lengthy and convoluted, can produce good results if only the decision-makers would set priorities and make the hard choices. In order to get the long-term future onto the agenda and incentivize the decision-makers to grapple with the hard choices, it will be necessary to give them clear, compelling rationales for action, champion these rationales inside and outside of government, and nurture the centers of excellence that will demonstrate the ability to act on the rationales. Chapter 7 examines the rationales for the exploration and development of space.

It should be clear by now that the numerous books and articles over the past three decades that blamed NASA for all of the space program's shortcomings in strategic vision and program execution didn't come close to telling the whole story. Although NASA bears some responsibility for flaws in management and decision-making, there's plenty of blame to go around. Bureaucratic organizations are susceptible to the same incentives that drive their executive and legislative masters. But they're also the locus of the government's technical expertise and institutional memory, which means they have to be an important part of the solution. Next, we'll consider the unorthodox notion that the bureaucracy—that much-maligned segment of the government—may hold part of the cure for our national nearsightedness.

Chapter 4
The Bureaucracy: Best Hope for the Future?

> *I am not an advocate for frequent changes in laws and constitutions, but laws and institutions must go hand in hand with the progress of the human mind. As that becomes more developed, more enlightened, as new discoveries are made, new truths discovered and manners and opinions change, with the change of circumstances, institutions must advance also to keep pace with the times.*—Thomas Jefferson (inscribed on the Jefferson Memorial in Washington, DC)

Bureaucrats as good guys

A favorite tactic of politicians who are campaigning for the presidency or a seat in Congress is to run against Washington. The nation's capital is portrayed as a monolithic beast that must be tamed. We hear this in every election cycle, particularly from newcomers aspiring to national office, but also from Washington veterans who want to move to higher office so they can "change the culture." Much of this rhetoric is aimed at partisanship and the gridlock it causes, but there is another aspect of Washington culture that often elicits the disdain of politicians and voters: the underperforming bureaucracy. For purposes of shaping the long-term future, however, is the bureaucracy really underperforming, or is it the government's best repository of futurist thinking, awaiting an incentive structure that allows it to unleash its creativity? Maybe "bureaucracy" isn't a dirty word after all.

Decision-makers can promulgate all the policies, pass all the laws, and start all the programs they want, but if there's no one to carry them out, nothing will happen. Bureaucracies are the implementers, the regulators, the detail planners, the builders, the operators, and the interface with the public. They also represent the institutional memory of the government. A congress-member's time in power can be short if they are not fortunate enough to get re-elected repeatedly. Presidents are limited to two four-year terms. The approximately 1,400 presidentially appointed positions in the executive branch are short-timers as well. In contrast, career bureaucrats can stay around for decades, building experience and making sure that expertise and historical perspective are not lost with each election.

Of course, this longevity can be either good news or bad news, depending on the characteristics of the worker. The stereotype of federal bureaucrats depicts workers who are, to put it kindly, not among the best and brightest, and not enthused about their work. They allegedly have a tendency toward excessive tardiness and absenteeism, and expect that they can never be fired unless they're convicted of a felony. Not the kind of person you want to have working for you, or managing your tax dollars, for their entire career. And certainly not the type of person you'd entrust with planning the nation's long-term future.

Fortunately, the people in the federal bureaucracy who match this stereotype are rare. Based on my own observations as an academic and experience as a government contractor, I've found federal workers to be well-educated (a high percentage of college and graduate degrees) and hardworking (routinely putting in more than 40 hours a week—sometimes much more). The same can be said for the contractors who work side-by-side with them. Admittedly, my Washington experience does not constitute a comprehensive survey of federal agencies. Since the mid-1990s, I've worked with the people who support the nation's efforts in space and defense at the Departments of Defense, State, Commerce, and Transportation, and at NASA. Their dedication is what I wish the rest of the country could see every time I hear unsubstantiated complaints about lazy civil servants wasting our tax dollars. When I look at government offices, more often than not I see programs that are *under*-staffed and *under*-funded, not the bloated bureaucracies of legend.

The government undoubtedly has its share of people who are shortsighted or unmotivated, and they seem to be the ones who get all the publicity. Of course, the private sector and other segments of society have a similar share of such people, but since our taxes aren't paying their salaries, you don't hear as much about them until they really screw up. Government bureaucrats, because they are paid by tax dollars, are held to a higher standard in which even a tiny percentage of failures or inefficiencies is considered intolerable, branding an entire class of workers as inferior. Americans should consider another perspective before they rush to judgment, and a good way to do that would be to read *The Case for Bureaucracy*, in which Charles Goodsell, a Virginia Polytechnic public administration professor, demonstrates that U.S. government bureaucrats get it right an overwhelming majority of the time, and match or outperform their counterparts in the U.S. private sector and in other industrialized nations.

As Goodsell points out in his book, bureaucrats in a variety of fields have proven they can be innovators. The biggest roadblock isn't the caliber of the people, it's the work environment and incentive structure in a hierarchical system based on several decades of industrial age and Cold War thinking. Such a system often struggles to keep up with the 21st century world, and it can't be changed overnight. There's no consensus on how the system should be modified to make it more vibrant, efficient, and responsive, as well as more attractive to new generations of potential government workers. Entrenched interests and conflicting ideologies are formidable barriers to anything more than incremental change.

Needed: opportunities to excel

Often, politicians and pundits get bogged down in arguments over whether or not the government is too big. This oversimplifies and polarizes the issue of the government's proper role to the detriment of attempts to address it. I believe most Americans don't know or care exactly how big the government is, or how it compares in relative size to the governments of other industrialized nations. They just want to get the best value for their tax dollars. We're actually doing better than many people realize. Our taxes are lower than in most comparable nations, even accounting for taxes levied at the federal, state, and local levels. And the U.S. bureaucracy, relative to the size of the country, looks pretty lean compared to foreign countries with a similar standard of living. Unfortunately, the percentage of Americans who believe the government is making wise choices and delivering good value for their tax dollars dropped significantly during the era of Vietnam and Watergate and has not recovered to prior levels. Increased political polarization since that era has perpetuated an environment unfriendly to constructive reform.

Even if we can get past unproductive arguments and agree on changes needed to make government organizations more agile and innovative, we're still faced with the problem of how to transition to a new and improved system without causing major shortfalls in government performance along the way. The difficulty of making this transition was demonstrated recently in the creation and early evolution of the Department of Homeland Security (DHS). The debate over whether and how to reshape federal agencies to respond to the full range of homeland security threats began in the 1990s but became compressed in the aftermath of the September 11,

2001 terrorist attacks. The new department, a consolidation of 22 federal agencies, officially began operations in January 2003. But there's a lot more to this assimilation effort than just changing logos on the stationery. It's a multi-year endeavor, much like merging a group of companies, and it's still underway. Early in the process, DHS set its sights on terrorism, which is a potentially high-consequence but historically low-probability concern (even after 9/11). Natural disasters received inadequate attention, even though they are high-consequence, high-probability concerns that can affect large geographical areas. (The U.S. has a 100 percent chance of being hit by hurricanes, tornados, floods, and earthquakes.) It wasn't until after the failed response to Hurricanes Katrina and Rita in 2005—which certainly qualified as a major shortfall in government performance—that DHS began to remake itself into an "all hazards" agency.

The formation and evolution of DHS is a consolidation effort, not a new way of doing business. Its hierarchy is modeled after other federal agencies, especially the Department of Defense. The hierarchical system that bureaucrats must live with is designed to ensure two things: accountability (to taxpayers, via senior decision-makers and the Congress) and security, both of which are necessary but unfortunately tend to stifle creativity due to their rigidity. This can become a pathology that calls for inordinate amounts of paperwork at the middle and upper management levels, which is generously shared with lower-level bureaucrats who helplessly watch as it dominates their workload. The capacity for long-term, innovative thinking is displaced at all levels. Seeing this, the best and brightest of the next generation are discouraged from aspiring to a careers in public service.

Accountability and security are governed by laws, executive orders, national and departmental policies, and long-established practice. That means change tends to be slow and incremental, with continuity almost always trumping inventiveness. It would be extremely challenging to convert key components of the federal government into flatter organizations with incentive structures that promote innovation and a long-term view. It's not the number of people, but rather the number of layers that hinders these characteristics. Too often, the path to senior decision-makers is a long one with multiple points at which good ideas (or status reports or problem disclosures) can be watered down, delayed, or stopped. The management chain is actually capable of moving much faster, mainly when an information request comes from the top of the chain. In that situation, all of the intervening management links scramble to respond as quickly as possible, often by the close of business on the same day. This is disruptive to the workers down below, and doesn't always produce the best answers because many questions can't be properly researched and reported in one day.

Occasionally, a senior manager will initiate a forward-looking action in this way: "We need an innovative solution to Problem X. Let's assign the appropriate people to brainstorm this and come up with a proposal in [a relatively short period, such as a couple of days or weeks]." This approach, sometimes called a "tiger team," allows a handful of people with the right mix of knowledge and experience to get creative while stimulating them to respond quickly. But often, that's not what will happen. The middle managers will set up a working group consisting of 25 or more people from every corner of the agency. The group will schedule too many meetings, at which people will spend most of the time talking about definitions, procedures, and management issues—sometimes referred to as "admiring the problem"—and eventually will ask for an extension of the reporting deadline. By the time the group reaches consensus and produces a report written by too many authors, it realizes that the problem could have been resolved weeks ago with a one-hour briefing to the inquiring senior official, presented by representatives from the one office that had the answers all along but was never invited to participate.

It may seem counterintuitive to make tasks more complicated than they need to be. But bureaucrats are sensitive to the same things that drive senior government executives and the Congress, who are, after all, their bosses. Those drivers, as noted in the previous chapter, are the electoral cycle, the budget cycle, and the news cycle. Government workers are not elected officials, but they have their own perspectives on how these cycles affect them. As is the case with the senior decision-makers, one of the effects is to discourage long-term thinking.

Like doctors who order extra medical tests to make sure they don't get sued for missing something, bureaucrats feel pressure—sometimes mandated by laws and regulations—to cover all foreseeable contingencies to avoid negative consequences to their agency, their division, their program, or their own job. The budget cycle is the strongest motivator, since agencies must justify their programs and request funding annually. But the election and news cycles are influential as well, especially for highly visible programs and favored projects of powerful politicians. Bureaucrats may take actions with a number of key questions in mind: Does the president or department secretary give personal attention to this issue? What needs to be done to ensure that top management submits an adequate budget request for next year? How will that request be received in the authorization and appropriations committees on Capitol Hill? Will there be cuts, or earmarks for the pet projects of congress-members? What will happen if the administration or key positions in Congress change as a result of the next election? Will agency projects attract the attention of the news media, and if so, will there be positive or negative effects on budgets or electoral fortunes?

The creative energies of hardworking people can be quickly dissipated in the presence of such concerns, coupled with other characteristics of bureaucratic environments:

- Large-group dynamics that stifle individual participation and enforce conformity.
- Constantly shifting of priorities.
- An unyielding atmosphere of urgency, including excessive use of emergency task assignments.
- Too many overtime hours, almost every week, and little or no flexibility to manage the workload.

Some people say that Washington is just like Hollywood, except the celebrities aren't as attractive. If that were true, all of us unattractive Washington types could shift to a California-style work environment like the ones offered to employees at companies like Google and Pixar, an atmosphere specifically designed to keep employees' creative juices flowing. But Washington isn't just like Hollywood, or even Silicon Valley, and it isn't because of the physical appearance, fashion, or fame of the people in power. The nature of the work carries with it responsibilities beyond those of any private-sector entity, and it requires a different type of creativity. Nevertheless, it would be helpful to cultivate a work environment that reorganizes the Washington bureaucracy and its vast community of support contractors to achieve something a little more like life in the Googleplex.

Of course, reorganization is not a magic elixir that cures all ills. Even if government bureaucracies succeed in reshaping themselves in beneficial ways, the nature of the work doesn't change. Lacking a profit-maximization goal, government workers often are chastised for having little or no incentive to control costs. But they probably wish their mandate was as simple, and their successes as clear and quantifiable, as those of the private sector. Rules designed to achieve particular public policy objectives can interfere with other goals, such as efficiency, timeliness, and cost effectiveness, putting agency bureaucrats in a no-win situation where different groups—factions within both their constituents and their overseers—demand different outcomes. As Goodsell tells us, bureaucrats generally must be content with making good progress on problems rather than solving them, but this inconclusiveness is not due to timidity, weakness, or ineptitude. It's because the problems typically have two characteristics:

> First, they are of a mass rather than a confined nature—that is, changes must take place in many places, in many things, and in many people. Second, they are not of a type wherein change can be achieved through the workings of the market on an automatic self-interest basis.

These complications manifest themselves in the formulation as well as the implementation of policies and programs. Depending on the amount of controversy and on current events, it can take anywhere from a few weeks to a few years to complete actions like creating or revising a national or departmental policy, initiating or modifying a major program, promulgating new regulations, changing procurement practices, or conducting a substantial study. Examples of some lengthy deliberations related to space policy include the Bush administration's National Space Policy, signed on August 31, 2006, which was three years in the making; the U.S. Space Transportation Policy of December 21, 2004, which was delayed almost two years due to the reassessment that followed the loss of space shuttle *Columbia* on February 1, 2003; and the Department of Defense Space Policy of July 9, 1999, which took over two years even though the action took place within a single department.

Concurrence is required from all parties affected by changes in policy and practice. Any significant dissent will result, at a minimum, in the tweaking of a few words, and at a maximum, in sending the entire document back to the drawing board. Since continuity is valued more than innovation, new policies tend to resemble old policies as much as possible. This makes it easier to reach agreement, and gives participants some confidence that the new policy can be completed sometime before their career is over. For those portions of a policy that require consistency over the long term, this is a good thing. But inspired visions are not the most likely outcome of this process. Even policies that are touted as visions can be merely a repackaging of old ideas, as Chapter 2 pointed out in regard to the Bush administration's January 2004 space exploration policy.

Fortunately, the government's mechanisms for innovation are lubricated regularly through support from academic researchers, think tanks, and federally funded research and development centers. These groups have more liberty to think analytically beyond today's hot issues, and in some cases are specifically chartered to do so. This helps to overcome two characteristics that inevitably creep into the system: parochialism, especially among people who spend their entire careers in the confines of a single technical discipline or bureaucratic community, never really understanding (and in some cases, never really caring) how the rest of the world thinks; and in-box agenda-setting, which leaves no time to think about the big picture or the long term because every day is more than filled with items requiring immediate attention.

Differing views of the government's role in space

Casual observers may have the impression that the space community is one big, happy family where like-minded people share the same goals and generally agree on who should be doing what. But since the space community, depending on how one chooses to define it, is made up of many tens of thousands of human beings with widely divergent backgrounds, that idyllic image is far from the truth. In general, the community includes scientists, engineers, policy-makers, administrators, contractors, direct users of space systems, and academic and non-profit researchers in related fields. Each group has its own goals, its own requirements, and its own agenda. The resulting conflicts can become heated and often counterproductive.

Former defense secretary James Schlesinger is credited with illustrating the differences between the U.S. military services by contrasting their interpretations of the phrase "secure a building." To the Army, it means seize and protect it; to the Marines, attack and destroy it; to the Navy, turn off the lights and lock the door; and to the Air Force, sign a three-year lease with an option to buy. The statement may be facetious, but the message is clear: despite sharing the mission to organize, train, and equip military personnel to protect the nation, and although they're all components of the same Defense Department, the armed services are distinct entities whose perspectives won't always agree. The differences can be even more stark across the space community, which is usually divided into military, intelligence, civil, and commercial sectors. The discussion here will focus on the latter two, since they are the ones seeking, each in their own way, to shape the future of space exploration and development and link it to societal evolution.

After a half-century of spaceflight, the respective roles of the government and non-government sectors are still not well established in some key areas, for a couple of reasons. First, there are long-standing and often passionate philosophical differences on the government's role in technological and economic development in general, and space activities can't escape this argument. Second, space is still a new and evolving human activity, so it's far too soon for things to have settled into anything resembling steady-state operations.

Some individuals and organizations believe there is little or no role for the government in space development. They see agencies like NASA as obstacles rather than partners. The Space Frontier Foundation, a group founded in 1988 to advocate large-scale space settlement, has not been shy about pointing out what it believes to be the space agency's ill-advised policies and misguided programs. The Foundation believes that "free markets and free enterprise will become an unstoppable force in the irreversible settlement of this new frontier," according to its website. The libertarian Cato Institute, an even more ardent supporter of free market capitalism, goes further. For years, it has advocated shutting down NASA as the best course of action for advancing the development of space. Clearly, this is a "throw the baby out with the bathwater" approach that would essentially end space science and space-based Earth science in the U.S., close valuable research facilities, reduce the nation's already low level of investment in aeronautics, and drive scientific and technical talent overseas. Finding the government's most appropriate and advantageous role in space is a multidimensional issue that is harmed rather than helped by such a one-dimensional solution.

Although extreme views don't provide useful guidance, they occasionally highlight points that resonate with more moderate perspectives. Few would argue with the notion that NASA sees space activity through its own lens, which tends to focus on the way things have been done traditionally, and only very slowly embraces changes that other parts of the community may wish to accelerate.

Planning the future of U.S. space exploration and development goes beyond what NASA can do alone, beyond even what it could do in partnership with the nation's other space-related federal agencies. But the government does have roles to play. If challenging macro-engineering efforts in space are to become reality, the most likely roles for the government are the same ones at which it has done well in the past:

- Fund and/or perform basic research and development that is likely to be underfunded by the private sector.
- Establish or encourage the building of infrastructure, such as spaceports and tracking facilities.
- Become an early adopter of new products and services, helping to stimulate and stabilize the initial market.
- Enact and enforce regulations in areas such as worker safety, public safety, and environmental protection.

Some critics would judge these actions to be inappropriate intrusions on the market, seeing them as attempts to create an industrial policy for space that would distort the market by allowing government agencies to pick winners, and prevent the private sector from developing products attuned to market demands. However, history demonstrates that these actions, though not always successful, have brought immeasurably large benefits to the nation. In their absence, infrastructure elements like transportation and communications, and utilities like electricity and water, would have experienced development cycles that were much slower, less ambitious, and more disjointed, resulting in dramatically slower growth of the economy and possibly preventing the emergence of a large middle class in the United States.

The debate over industrial policy has raged since the end of World War II, when the government became a dominant player in the nation's research and development efforts. Should the government be involved in promoting particular industrial sectors, and if so, how much and by what means? If such involvement benefits society, producing "public goods" such as weather forecasting, it would seem to be a worthwhile investment of taxpayers' money. But there is no guarantee that the government will routinely make the best choices about where to invest. Bad choices can sink large sums of public money into efforts that produce no benefits at all, or which give an unfair advantage to a select group of profit-seekers in the private sector.

On the other hand, there is no guarantee that market forces will lead to the best investment choices, especially where societal benefits are concerned. When profitability is highly uncertain or seems to be too far down the road, private sector entities are disinclined to risk their near-term health on long-term probabilities, especially when they are unlikely to capture all the resulting benefits of their investments. For example, private companies would not be interested in taking over the vital responsibilities of the U.S. weather satellite fleet or the U.S. Coast Guard in the absence of large federal subsidies. Both were privatization targets during the Reagan administration.

Although governments tend to be unsuccessful at "picking winners" for commercial competition, encouragement of space commerce is mandated in national policy so the government must pick something. The methods by which the government could aid space commerce are many and diverse: targeted research programs, direct subsidies, patent licensing, regulatory or anti-trust relief, tax breaks, low-interest loans or loan guarantees, provision of infrastructure, guaranteed government purchases, and liability indemnification.

Unfortunately, space development, along with other technology and trade issues, often gets mired in partisan debates where liberals, moderates, and conservatives disagree on the amount and type of government intervention. This has resulted in policies that are uncoordinated with each other and inconsistent over time. This is particularly unwelcome in a period when the space capabilities and marketing aggressiveness of international competitors is growing.

A decision by a government not to have an industrial policy is, in effect, a decision to have a haphazard industrial policy. Realities such as population growth, technological advances, economic downturns, foreign competition, natural disasters, and national security compel governments to take many industrial policy actions, as the United States has done by investing in research, maintaining a federal highway system, operating an air traffic control system, bailing out critical industries, subsidizing education, providing job retraining programs, furnishing subsidies to farmers, granting tax breaks to corporations, and entering into sole-source contracts for some government procurements. Coordinating these actions in support of top-level, long-term goals is a far more sensible practice than initiating them randomly, which inevitably leads to conflicting program objectives. To dismiss this as a "planned economy" approach would be inaccurate and shortsighted. It's not socialism, it's common sense: choosing to be proactive rather than reactive.

There are many examples of great benefits delivered more quickly to society as a result of the coordinated efforts of the public and private sectors. The railroads of the mid-19th century were built with private investment, but their rapid expansion and initial economic viability depended on the federal government's grant to the rail companies of the right-of-way for laying their tracks. In the early 20th century, airline companies established themselves with the help of mail delivery contracts from the U.S. Post Office. Also, air travelers should ponder how much their tickets would cost if the airlines had to own, operate, and upgrade the air traffic control system. They may decide to stick with ground transportation—but that would be more expensive too. If all highways were privately owned, they would all be toll roads, and would only exist in areas where a profitable amount of traffic was envisioned. That would alter regional development plans in a multitude of ways. It would also increase the cost of doing business for the trucking industry, which means higher prices for most of what we buy.

The satellite communications industry is an excellent example of the stark difference between the government rejecting versus embracing involvement with the private sector. We can compare the results of the two approaches because the government followed a hands-off policy in the Dwight Eisenhower administration, then reversed course and became a generous sponsor and partner of the industry in the John F. Kennedy administration. Corporate interest in communications satellites existed in the U.S. in the 1950s in companies like RCA, Hughes, and AT&T, all of which worked on designs for orbiting relays. Government support for design research was not forthcoming from the Eisenhower administration, but more importantly, the government refused to provide launches to the private sector. The companies shelved their plans because they had no intention of investing in their own rockets and launch pads at that time. However, starting in 1961, Kennedy supported fast-paced development of a global geostationary constellation and the U.S. Congress agreed. NASA conducted research on communications satellites and launched commercial payloads, allowing high-orbit geostationary relays to become reality starting in the mid-1960s. How long would this development, with its profound benefits to the world economy and foreign relations, have been delayed if the Eisenhower policy had prevailed? Since we know that the first successful commercial launch services in the U.S. didn't appear until the end of the 1980s, it's not unreasonable to assume a delay of a quarter-century in the emergence of commercial satellite communications in geosynchronous orbit—unless foreign ventures achieved it more quickly.

Exploring and developing the solar system while bringing benefits back to Earth is a bigger job than either governments or the private sector can do alone. Mutually supportive policies, investments, and programs are required across all the players. Learning from past experience is essential, but so is recognizing and accepting that the future is likely to be very different from the past. The rest of this chapter explores what this means for NASA, and the next chapter does the same for the commercial space sector.

NASA's evolving role

When the space age began, NASA appropriately took on a broad range of responsibilities because there was no one else at the time that could. Fascination with NASA in the 1960s had more behind it than just Project Apollo, which had the highest visibility. Concurrent with Apollo, NASA was working on launch technologies, communications satellites, weather satellites, deep space probes, and an assortment of scientific disciplines, some traditional and some newly created for the study of space. The space community, consisting of government, business, academia, and non-profit institutions, was far smaller than it is today, but results came quickly due to a variety of factors: political and funding support, the perceived urgency of the mission, the newness and excitement of the enterprise, and the numerous projects of the early space age that could target "low-hanging fruit" and yield near-term results. In the decades since then, the pioneering work of the space agency has enabled other parts of the community to grow so that key functions, even very challenging ones, can be handed off to organizations outside of NASA. But the redistribution of tasks is far from complete.

Like any organization, NASA wants to ensure its continuation as a productive entity. It doesn't want to shrink dramatically in personnel or budget, and it doesn't want to give up programs that it sees as its strong suit. Critics, on the other hand, assert that it must give up some programs as the space community evolves, particularly those that involve designing, building, and operating launch systems and orbiting platforms (other than science satellites). It's not surprising that NASA would resist calls to back away from its work on major hardware development programs that it sees as its lifeblood, especially those connected with human spaceflight. Shifting national priorities and tight budgets have injected an undercurrent of existential threat into the agency's culture since the end of Apollo.

NASA, despite its high visibility and historical achievements, is an independent agency that has limited clout in the grand scheme of American government. It's not represented by a Cabinet secretary, unlike other space players in the government: the Defense Department, the State Department, the National Oceanic and Atmospheric Administration (NOAA) in the Commerce Department, and the Federal Aviation Administration (FAA) in the Transportation Department. NASA is not a particularly large agency, employing less than 20,000 civil servants. Its fiscal year 2008 budget of $17.3 billion was less than the Defense Department spent on fuel during that year (approximately $20 billion).

Civil space policy has only rarely and briefly been high on the agenda of the president and key policy-makers since the end of Apollo. From the perspective of most decision-makers today, civilian space efforts have become decoupled from U.S. interests such as international security and the national economy. The essential unanswered question for the civilian space program has been: What is the space program's link to U.S. national interests? The answer is undoubtedly a moving target that will keep moving as the century progresses. Civil space since Apollo has been cast not as a strategic program, but as an issue combining science and technology policy (which is remarkably low on the agenda despite its importance to society), domestic job creation, and potential foreign policy benefits.

After its victory in the Moon race of the 1960s, the NASA of the 1970s and 1980s was an organization struggling to deal with its post-Apollo identity crisis. The 1990s brought the end of the Cold War, which rearranged the playing field in the space agency's search for its place on the national agenda. The current era, dominated by issues such as homeland security and the global economy, further complicates efforts to keep the civil space program from being politically marginalized.

NASA will continue to evolve, as an organization and as a component of the executive branch, in the decades ahead just as we've seen it evolve in the past half-century. But will its future be determined by long-term strategic planning (choice) or by short-term circumstances and budgets (fate) unrelated to a coherent approach to space exploration and development? By now it should be clear that a short-term approach currently dominates, and it will take a strategic vision—a real one, not a repackaged mid-20th century vision—and active intervention to create an incentive structure that will bring out the best in the NASA bureaucracy.

If we continue conducting business as usual, changing circumstances in the coming decades could result in realignment of NASA's priorities, including a diminished role for human spaceflight. For example, NASA's research and development prowess could be redirected away from space missions to address the challenging issues of the day. This has already happened in the agency's history. The "energy crisis" of the 1970s prompted amendments to NASA's enabling statute, directing the agency to work on solar heating and cooling (added by the Solar Heating and Cooling Demonstration Act of 1974) and "ground propulsion technologies" (added by the Electric and Hybrid Vehicle Research, Development, and Demonstration Act of 1976). At that time, NASA also was working with the Energy Department on wind turbines for power generation. Similar pressures to divert NASA resources and personnel can be anticipated in the future, because technological issues such as energy and transportation have become more complicated and more urgent.

Expectations of terrorist threats, especially using weapons of mass destruction on the U.S. homeland, could result in requests from the Department of Homeland Security or law enforcement agencies for NASA assistance. DHS is still in the early stages of determining how space systems and space-derived technologies can be applied to its mission. DHS already has consulted NASA on the subject of robotics, and eventually will need to increase its sophistication in the use of remote sensing and geographic information systems. DHS interest in unmanned aerial vehicles for surveillance is well known, so it's not unreasonable to anticipate future interest in the operation and applications of satellite systems. Spin-offs from space technologies may help DHS develop methods to detect contraband using various kinds of sensor systems at airports, seaports, and border crossings.

Since the 1980s, we've seen Earth science become a mainstream mission area for NASA. If environmental and climate change problems grow worse and public concern increases, NASA will be one of the first organizations the nation looks to for information and solutions. When that happens, there will be an expectation that NASA has maintained a high level of activity and expertise in this area. The agency must not leave the perception that it has sacrificed Earth sciences for the sake of human spaceflight ambitions, to the detriment of down-to-Earth societal needs. The Obama administration and recent Congresses have been supportive of Earth science, so this work doesn't appear to be in political jeopardy in the near future. However, sustainment of this line of research at the level necessary is proving to be a serious fiscal challenge.

Across the interested American public, opinions on the best space-related uses of tax dollars will vary over time and among interest groups. Some will endorse the traditional view of NASA as a space science and human spaceflight agency. Others will prefer redirection of NASA efforts toward a primary focus on Earth science or ground-based technology development. A strong case can be made for each of these positions, but they should not be allowed to cause gridlock in the essential task of formulating a long-term strategy and specifying the milestones needed to carry it out. I believe this debate can be resolved without pitting NASA's mission areas against one another, and without making a stark choice between a program aimed inward toward the Earth or outward toward space. The critical change needed to make this happen is that NASA must transition its culture from "follow us and we will lead you to the stars" to "we will enable you to go to the stars."

NASA is still in the business of building and operating space infrastructure, primarily launch systems and habitable spacecraft. With the space shuttle decommissioned and the space station complete, the agency plans to reorient itself to a mostly developmental mode as it works on its next-generation launch vehicles and prepares the systems that will be needed to return to the Moon and set up research facilities there. After a few years of development, NASA expects to shift back into operational mode with its new transportation system and the lunar infrastructure elements that emerge from the process. This is the point at which current planning takes a wrong turn. NASA should not be looking forward to returning to operational functions that will once again overwhelm its efforts and resources. Indeed, the 2009 Augustine Committee report mentioned in Chapter 2 identified the fiscal competition between development and operations as the "fundamental conundrum" of the NASA budget, virtually guaranteeing programmatic gaps in the absence of large budget increases. Instead of perpetuating this situation, the agency should be eagerly working toward a time, as soon as possible, when it can wash its hands of operational duties and concentrate on what it does best.

NASA's history demonstrates that the agency's greatest successes, even in the wake of serious setbacks, have been on programs with limited, well-defined objectives and finite durations (for example, Apollo, Skylab, and numerous space science missions). On the other hand, open-ended programs with less clearly defined or too many objectives have proven problematic, undermining their political and economic sustainability (such as the space shuttle, International Space Station, and various attempts at next-generation space transportation systems). Solar system exploration and development is an open-ended program of incomprehensively large proportions, signaling the need for a rethinking of the NASA's traditional leadership and management model.

NASA should continue to operate its clearly defined science missions, including those with long durations such as space telescopes and deep space probes. At the same time, we're reaching the point where NASA will be able to take big infrastructure operations tasks off of its "To Do" list and hand them over to the private sector and to operational agencies of the government. This should not be viewed negatively, as somehow demoting the agency or disparaging its abilities. On the contrary, it should be considered a badge of honor, in keeping with practices that go back to the earliest days of the space agency when two of the most successful space applications ever devised were moved to new operators as soon as they reached the appropriate level of readiness. As already noted, satellite communications became a highly successful private-sector activity. Also, weather satellites were intended from the beginning to be operated by the government's weather service (although NASA still handles procurement of civilian weather satellites) and the Defense Department. In addition to having operational goals from their inception, another common factor contributing to the spectacular success of communications and weather satellites is that they served a clear, immediate need and were able to quickly mesh with the terrestrial elements of the functions they served.

Other space applications developed within the government, such as launch services and satellite remote sensing, didn't have as clear a path to follow, so their transition to private operation has taken longer, been much more difficult, and continues to involve substantial subsidies. Although their path has been more challenging, demonstrating that space technologies can't all be "commercialized" or "operationalized" at a similar rate, the direction has been the same: away from the research and development organizations that spawned them, and toward operational entities that will expand and improve them.

Another example of maturing technology moving beyond the space agency is one that most people have never heard of: the NASA Electronics Research Center (ERC) in Cambridge, Massachusetts, which opened in 1964 and closed in 1970. In part, the ERC was a victim of waning budgets as Apollo-era spending declined, but closing this center also made strategic sense for NASA. It already had become evident that U.S. companies large and small were investing heavily and would surge ahead in areas such as computers, electronic components, data processing, information display devices, microwaves, and lasers, so NASA didn't need to be concerned that the private sector would under-invest in these technologies. Taxpayer resources could be directed elsewhere.

Transfer of information and technology to the wider community has been part of NASA's charter since its beginning. Transfer of space systems and infrastructure responsibilities to new operators also should be seen as fulfilling that mandate. Space capabilities have become widely distributed, so NASA doesn't need to retain responsibilities that others can handle. As the agency moves beyond its current programs it can be relieved of the burden of spending a third of its budget, as it has in recent years, operating launchers and a space station. This change won't be implemented overnight—NASA has long-term obligations to its international partners on the space station, for example—but the transition should begin and can be accelerated when appropriate, freeing NASA to focus on the science, exploration, and technology development tasks at which it excels.

Moving beyond space infrastructure operations currently is not part of the strategic plan for the space agency. Instead of welcoming the opportunity to cast off this workload and focus its total resources on cutting-edge research and development, NASA still clings to operational duties and has taken only small steps to shed 20th century responsibilities that now or in the near future can be left to others. The agency need not fear that it won't have exciting things to do if it gives up routine space operations. Robotic exploration of the solar system will continue, and will become increasingly sophisticated. As we'll discuss in a later chapter, there's also the little matter of helping to save the planet using space technology. Aside from that, the private sector will still need its government partner to help buy down the risk in ventures like:

- Improving the affordability, reliability, flexibility, and safety of access to space.
- Advancing the state of the art in space robotics.
- Building, maintaining, upgrading, and recycling space hardware for larger, more capable facilities in space.

- Controlling the space debris problem, and creating a used satellite market in the process.
- Creating off-planet industries to take advantage of the microgravity environment, ready access to hard vacuum, and the vast material and energy resources of space.

The new paradigm of NASA as an enabler has much in common with the agency's predecessor, the National Advisory Committee for Aeronautics (NACA), which provided research facilities and expertise to the nation's growing aviation industry for over four decades. NACA did not build and operate a fleet of aircraft or a network of airports as part of the national aviation infrastructure. Those responsibilities were left to other government agencies and the private sector. Similarly, the mature NASA of the 21st century should operate its own systems only for scientific missions and engineering research, and leave it to others to provide products, services, and infrastructure.

This transition can't be successfully accomplished in an environment where the NASA bureaucracy perceives that its survival is threatened by a loss of capabilities or mission. Actions that appear to be little more than government downsizing and budget-cutting will be disincentives to long-term thinking and will undermine attempts at strategic planning. Civil servants will be discouraged from becoming farsighted technical and programmatic innovators if their leadership seems enslaved by short-term budget cycles, or their president has a "government is the problem" attitude, as was the case for much of the past three decades.

The NASA leadership recently took an action that demonstrated the predominance of short-term planning and sent a signal to its employees and to the broader community that the agency was unprepared to contemplate its future beyond currently funded programs. In August 2007, the NASA Institute for Advanced Concepts (NIAC) ceased operations. Based in Atlanta, Georgia, NIAC had been established in February 1998 to offer small grants to universities and businesses studying technologies that would leap ahead of current capabilities and be attainable in 10 to 40 years—precisely the timeframe that should be central to NASA's strategic planning. The staff of six people (not all of them full time) processed 1,309 research proposals during NIAC's nine-and-a-half-year existence, and awarded 126 Phase I grants and 42 Phase II contracts for a total value of $27.3 million. NIAC was costing NASA an average of $4 million per year for grant awards and its small staff. Through this investment, the agency was discovering new scientific and technical talent and encouraging concept development that was useful in planning for tomorrow's challenges.

NIAC was shut down so that its meager funding could be redirected to the inaccurately named Vision for Space Exploration, one of the agency's largest programs. Ironically, NIAC's low-budget research effort, designed to look ahead to mid-century, was sacrificed to give a tiny funding increment to a program that lacks a strategic purpose or plan beyond a proposed return to the Moon in 2020. Apparently, no one recognized that NIAC was exactly what the space exploration program needed, which could have been obtained by linking NIAC more closely to exploration efforts and giving it more funding rather than terminating it.

The closing of NIAC provides evidence that the agency continues to be in survival mode, more concerned about shoring up this year's and next year's budgets than with crafting a long-term strategy for science, technology, and exploration in the national interest. This observation should not be interpreted as demonizing NASA, as so many critics have been fond of doing in recent years. It should be seen in the context of the discussion so far in this book: although NASA management cannot be absolved from responsibility for its own decisions, the agency is operating in a larger environment that encourages near-term expediency at the expense of long-term strategic thinking. NASA's talented civil servants deserve to be set free from this shortsighted survival mode. They need room to innovate and inspire through the pursuit of well-considered goals that make a difference today and will continue to do so a half-century from now.

Chapter 5
Astropreneurs: The Real Vision, or Just a Dream With Good Special Effects?

> *In spite of the opinions of certain narrow-minded people who would shut up the human race upon this globe . . . we shall one day travel to the Moon, the planets, and the stars with the same facility, rapidity, and certainty as we now make the ocean voyage from Liverpool to New York.*—Jules Verne, *From the Earth to the Moon*, 1865

> *We want to build colonies on the Moon, Mars, the moons of other planets, and even nearby asteroids. We want to make space tourism and commerce routine.*—Daniel S. Goldin, NASA Administrator, 2000

Space entrepreneurialism is not a new development—it's been around at least since the beginning of the space age in both the real world and in science fiction. For decades, there have been people who see outer space as more than just a playground for scientific and engineering research or a "high ground" for military and intelligence interests. They see it as a place ripe for economic development, an industrial park of astronomical proportions.

Many of today's ambitious entrants to the space business, especially those offering alternative launch concepts, like to refer to themselves as NewSpace, as if they were part of some social movement. In a way, they are, because they're purposefully trying to distinguish themselves from government space activities and the traditional contractors that serve the government. I've never been enamored of the NewSpace label. It seems inappropriate to call something "new" for very long, unless it's a piece of real estate like New York or New Jersey. At some point, this movement can't claim to be new anymore, and they'll want to distance themselves from the negative connotations of newness, such as inexperience. I prefer the more colorful term of astropreneur—a developer, and often an evangelist, of space commerce.

The terms "space commerce" and "commercialization of space" have been used to describe a variety of activities. For our purposes, these concepts are defined as the provision of products or services by the private sector that are directly dependent on one or more space systems and have the following characteristics:

- Private capital is at risk in development, upgrading, and/or operation.
- Primary responsibility and management resides with the private sector.
- There are existing or potential non-governmental customers.
- Market forces, such as demand and competition, ultimately determine viability.

Framing the issue in this way doesn't exclude the possibility of government investment, even to the extent of public-private partnerships, or of government agencies as part of the market. This definition also recognizes space commerce ventures that result from privatization, which is the transfer of a publicly funded system, originally designed to be run by the government, to a private-sector owner or operator.

In some ways, outer space is fundamentally different from other endeavors in the business world. Space is exceedingly hard to get to, and once you put your space segment in place the physical environment is unrelentingly hostile. It's extraordinarily difficult to maintain space assets because you can't retrieve them or send a repair crew (at least not currently). The legal and regulatory regime for space differs in many ways from comparable activities on the ground. For example, there is no private property system for extraterrestrial real estate, and the international law governing salvage in space is more restrictive than that which applies to salvage at sea.

In other ways, outer space is an environment much like terrestrial business endeavors. Economic viability still depends on the same factors that drive comparable Earth-based businesses. Opportunities may seem endless in a realm that provides ideal vantage points for observation and signal relay, and where raw materials and energy are essentially unlimited. But success will still require making the best use of resources and human capital, and investors' expectations for timely and reasonable returns on their investments remain a critical driver. In the investment community, the glamour of space attracts some, but others are repelled by the inescapable realities of space projects: high cost, high risk, and long payback periods. Even daring venture capitalists are taken aback by proposals that are likely to take more than a decade to start paying off, if they pay off at all. So the potential payoff had better be big.

How big is space—as a business venture?

According to the Federal Aviation Administration's Office of Commercial Space Transportation, private-sector space revenues exceeded government space expenditures for the first time in 1997, both in the U.S. and worldwide. Also in that year, the number of commercial payloads launched into space exceeded the number of government payloads.

The Space Report 2009, an annual survey by the Space Foundation of Colorado Springs, Colorado, found that global space activity in 2008 had totaled $257 billion, an increase of $6 billion from the survey's estimate for the previous year. The *Space Report* attempts to tally all government and commercial expenditures in a variety of space sectors including satellite manufacturing, satellite services, launch services, ground equipment, and other space infrastructure support. Similarly, the Satellite Industry Association (SIA) in Washington, DC puts out an annual report covering the same sectors, but uses a somewhat different methodology that excludes non-commercial government spending. SIA's total for 2008 was $144 billion and is probably a better indicator of the level of space activity from a commercial perspective. Perhaps the most impressive finding from SIA was that space commerce has had a long stretch of double-digit growth, averaging 14.2 percent annually from 2003 to 2008.

Space commerce has proven remarkably resilient in the face of downturns in the world economy, and is expected to remain so as long as economic slumps don't stretch for more than a couple of years. But despite continued growth, space industry stock values were hit hard along with everything else in the recession of 2008-2009. The illogic of the situation was illustrated in remarks made at a December 2008 conference by the chief financial officer of a satellite services company. He noted that despite good financial performance, his company's stock had sunk so low that its book value was $30 million less than the cash the company had in the bank.

If we accept SIA's estimates of revenue and growth, and assume that stock values will rebound, then it looks like space commerce has become quite large and successful, making it part of the mainstream of the world market. Or is it? Measures of size and success are relative to their environment. The top seven companies on the 2009 Fortune 500 list *each* had revenue in excess of SIA's estimate for the entire space industry. The top three companies on the list, Exxon Mobil, Wal-Mart, and Chevron, each had 2008 revenue—$443 billion, $405 billion, and $263 billion, respectively—that surpassed even the higher estimate of the Space Foundation, which included the non-commercial spending of governments. It seems inconceivable that individual companies could pull in more money each year than the whole world spends on all types of space activity. Exxon Mobil's 2008 *profit* of $45 billion was two and a half times NASA's budget for that year. One could fantasize a scenario in which several of these companies pool their resources and start their own space program, quickly outstripping everyone else and eventually ruling the cosmos.

Before getting too depressed about the financial insignificance of space commerce, it's important to realize that revenue totals don't tell the whole story. We also need to consider how many non-space industries are enabled or significantly aided by space capabilities. If we tried to list every industry that depends on weather forecasts, long-distance communications, precision navigation and timing, and remote observation, it would be a very long list. Terrestrial alternatives exist for all of these functions, but space systems are a critical part of the mix that vastly improves characteristics such as accuracy, affordability, and availability. National security organizations and think tanks have conducted seminars with the theme "A Day Without Space," incorporating military, intelligence, and economic considerations, and the findings emphasize that all sectors of modern society have become reliant on space, and would have a very bad day indeed if space capabilities suddenly disappeared.

Although space commerce is ubiquitous and growing, few people are aware of it because space services are highly reliable and mostly invisible. Generally, this is a good thing. Any attempt to conceptualize the perfect infrastructure for delivering a service would certainly envision reliability and invisibility (or at least, unobtrusiveness) as essential characteristics. People don't want to think about whether their phone call is being carried by satellite, fiber optic cable, or tin cans connected by string; they just want the call to go through properly. Most people don't marvel at the array of cool space technologies that are used to bring them the daily weather forecasts. And who gives a thought to the multitude of retailers that verify our credit card transactions in real time using satellite links? The downside of this transparency is that people who routinely derive benefits from space services, but who don't realize it, may be reluctant to support government or private investment in space systems that are very capital-intensive and may be inaccurately perceived as catering to a small, privileged user community someplace far from themselves.

Consumer recognition of space services is slowly awakening as more people subscribe to satellite television and radio, or use Global Positioning System (GPS) navigation units with an awareness that signals from multiple satellites are involved. This boost in recognition is welcome, since people need to have an appreciation for what space commerce has been accomplishing before they'll be willing to invest in the next generation of space services.

Risks and (sometimes) rewards of today's space commerce

The first step in strategizing about how to create the new and improved space commerce of the 21st century is to review the lessons learned from what's occurred so far. What has been successful, and why? What ventures failed or struggled mightily, and why? Which characteristics were unique to each venture, and which can be generalized across all commercial space efforts? Books filled with case studies could be written about these questions, tracing the abundance of good and bad experiences going back to the earliest days of the space age. For our purposes, we'll just take a brief look at where we are and how we got here.

Success in space businesses has been demonstrated when: 1) a service in common use can be performed as well or better from space, 2) the space capability adds capacity needed by the user community, and 3) its price is competitive with terrestrial alternatives. This can come in the form of space systems that are privately owned (communications satellites), government-owned systems that generate a value-added market (weather monitoring and navigation), or a combination of both (remote sensing).

Satellite communications is the classic success story of space commerce, and the reasons for this are clear. Communication is an inherent part of almost all human activity, so better communication—faster, easier, more affordable, more flexible, more extensive—is always a marketable idea to better serve businesses, government agencies, vehicles, and individuals. People are eager to communicate with each other, and to pay for news and entertainment, even in situations where they're not making money from it. Each time in human history that better communication techniques were developed, they quickly gained wide acceptance, not only improving existing services but also creating new ones. Orbiting satellites are another step in that continuing evolution characterized by huge leaps in capability and substantial reductions in user costs.

In addition to its universal demand, satellite communications had other advantages that made its success all but inevitable. Most of the communications infrastructure, both physical and regulatory, was already in place when satellites appeared. Once satellites were launched and ground stations were built, it was a relatively simple matter to plug them into existing networks, particularly in developed countries where large numbers of customers were ready to take advantage of the new services and additional capacity.

As noted in the previous chapter, the promise of a global satellite communications network was promulgated with great fanfare by the Kennedy administration, and backed up with enabling legislation and solid funding for research and development. Strong government support and public awareness are a powerful combination, especially when ongoing economic benefits are visible. This put satellite communications on its road to success, with just a few bumps along the way that the industry has been able to overcome.

The success of satellite navigation played out similarly starting in the 1990s because it too offered valuable services to an extensive global market, and was strongly supported by U.S. policy. In this case, the U.S. government owns and operates the GPS satellite constellation, offering the signals free to users around the world. New navigation satellite systems from Europe and Asia are joining the market, which should ensure its continued growth.

Space applications other than communications and navigation have followed a rockier path where inconsistent government policies, inadequate funding, and unrealistic business plans have inhibited the emergence of fledgling industries. Remote sensing illustrates these problems in a long, convoluted story that goes back to the mid-1960s. Imagery of the Earth from space has a large array of applications—for example, in mapping, urban planning, resource exploration, environmental stewardship, and agriculture—leading one to believe that it should enjoy success comparable to that of satellite communications. But the value of remote sensing is often difficult for the average person to comprehend since it requires special training to turn raw data into useful information. Some of those skills can be encoded in software and run on ever-more-powerful personal computers, but this was not the case for much of the history of remote sensing. Policy-makers typically have had no better understanding of its capabilities and potential than the average person, resulting in ill-conceived policies and unsteady federal support.

Remote sensing also lacked other advantages that sparked the rapid acceptance of satellite communications. While demand for overhead imagery is large, it's far from universal. The product itself is poorly defined, in sharp contrast to homogeneous services like telephone calls or TV broadcasts. The imagery can have applications for either private profit or public good, or sometimes both in the same image, making ownership and use issues a challenge in the legal and regulatory environment. Initially, no physical or regulatory infrastructure to aid implementation was in place to aid the emergence of civil remote sensing, and even today a fragmented community exists that sometimes acts in counterproductive ways. And with the exception of weather satellites in the early 1960s, no high-level fanfare heralded the appearance of remote sensing, and no long-term funding commitment was made by the U.S. government.

The global market for satellite remote sensing remains reliant on government subsidies and derives at least half of its revenue from government customers. Nevertheless, the number of commercial Earth imaging satellites built and launched during the next decade is expected to increase. Success will depend on the timeliness and affordability of images as well as their quality.

Launch services are another troubled area of space commerce. This may seem counterintuitive—the launch business should be a sure winner, since every other space business is absolutely dependent on it, and there are currently no alternatives to chemical rockets for getting payloads into space. In reality, financial success depends on having government space agencies as anchor customers. Even that is no guarantee of success. Ask anyone in the space community, "How do you make a small fortune in the space launch business?" and their answer will be, "Start with a large fortune."

Hauling satellites to orbit bears little resemblance to hauling freight by truck, rail, ship, or aircraft. The payloads are few in number, with the dominant part of the market being just 20 to 25 commercial satellites to geosynchronous orbit per year. The payloads are delicate and expensive, most of them costing well over $100 million. Despite decades of experience, launch costs remain stubbornly high at about $20,000 to $25,000 per kilogram to geosynchronous orbit. Launch schedules can be erratic because they are thrown out of whack by something as serious as a launch failure or as simple as delays in the supply of parts.

Comparing the timeliness and success rate of the space launch industry to its terrestrial cargo-carrying counterparts demonstrates what a tough business this is. Traditional shipping services deliver on a fairly predictable schedule, often book shipments on short notice, and hardly ever blow up the cargo. The same can't be said for space launch services. Delivery is frequently held up by technical glitches and bad weather, flights need to be booked years in advance, and reliability rates for large commercial launchers are about 95 percent. Imagine a trans-oceanic shipping company that promises to deliver your extremely valuable cargo to its destination in two years (a boat would be specially built for it and used only once), there's a one-in-twenty chance that your package would be destroyed along the way, and the freight charge is approximately 50 to 100 percent of the value of the cargo, plus insurance. Such terms would not attract many customers. Clearly, today's space transportation services shouldn't be measured by terrestrial standards, but if the grand plans of space visionaries and entrepreneurs are to be carried out, someday they will be.

The launch industry has another negative aspect: it's politically sensitive. This is so because the technology is applicable to military uses, having been built on the heritage of long-range ballistic missiles. Also, rocketry presents clear hazards to launch participants and uninvolved third parties, making it a public safety concern. These factors initially made the practice of launch operations a government responsibility. It wasn't until the 1980s that private-sector launch services made concrete steps toward viability due to industry lobbying and the combined efforts of the business-friendly (and privatization-friendly) Reagan administration, generally supportive congressional committees, and the years-long efforts of federal agencies to establish a suitable regulatory regime.

After decades of effort, the quest continues for affordable, reliable, flexible, frequent access to space. New U.S. commercial launchers capable of lifting large payloads to various orbits have yet to prove themselves and pose a challenge to the evolved versions of U.S. Air Force missiles of the 1950s and 60s, or to the growing list of foreign competitors. Why would astropreneurs choose to focus primarily on the seemingly unsolvable problem of better, cheaper access to space after numerous abortive government and private efforts? Do they like losing money and being embarrassed by failed launch attempts, or have they crafted plans that will open up the solar system for those willing to be patient with unusually long development cycles? Either they're crazy idealists, or they're canny visionaries willing to tolerate longer time horizons than most of the business community. Or maybe they're a little of both.

The birth of a business movement

Astropreneurs recognize that access to space is the number one concern of everyone interested in exploration and development. They also recognize that the private sector has the technical capability to contribute solutions. However, corporate interests cannot be confident that they will get back what they've invested within a reasonable timeframe, so they usually look to the government to share the risks. But not everyone sees this as a path to success.

The origin of this discontent with the traditional government-led approach goes back to the space advocacy movement that appeared in the U.S. in the aftermath of the Apollo program. At that time, the fortunes and budgets of civil space efforts declined dramatically. Shrinking or disappearing programs put the future of space exploration and development in doubt. Space advocates perceived a need for action to stave off decline, at least until the space shuttle would start flying, at which time, they believed, the rapid pace of space development would resume.

The movement flourished in the 1970s and 80s. Gradually, the numerous advocacy groups realized that the government was never going to return to the way things were in the thrilling days of the Moon race. This was confirmed in the aftermath of the 1986 space shuttle *Challenger* accident. Some groups began to question whether NASA would ever help them fulfill their dreams, even if the agency was given all the resources it needed for its shuttle and space station programs. The organizational and cultural weaknesses revealed in the *Challenger* investigation convinced some that lobbying on behalf of NASA was a wasted effort; rather, the focus should be on establishing a friendly environment for private-sector development of space.

Advocacy groups have sought to serve a variety of interests, such as space science and education, but economic ambitions have been especially prominent, including employment opportunities, industrial growth, and the development of new forms of space commerce like space tourism and settlement.

The previous chapter briefly introduced the Space Frontier Foundation, which formed in 1988 (not to be confused with the Space Foundation of Colorado Springs, mentioned earlier in this chapter). When it began, the group's founders articulated what they termed "three truths":

- Large-scale industrialization and settlement of the inner solar system is technically possible within one or two generations.
- This is not happening (and can't happen) under a centrally planned and exclusive U.S. government space program.
- The existing bureaucratic program must be replaced with an inclusive, entrepreneurial, frontier-opening enterprise, primarily by working on the outside to promote radical reform of U.S. space policy.

Concluding that no existing organization was appropriate to this task, the group's creators decided that the responsibility fell to them, and that the Internet would allow them to get their message out to the world undiluted. The Foundation's approach has included strongly worded and often antagonistic statements (in trade publications and posted on its website) that blame NASA and its major contractors for stifling entrepreneurial space commerce and human settlement goals. Like other space advocacy groups, the Foundation's website touts its perceived influence on space policies and programs. Realistically, the overall impact of U.S. space advocacy groups on major space policy and program decisions has been minimal. They have lent their voices in support of government space projects conceived without their input, sometimes claiming success and sometimes suffering failure in their attempts to boost the fortunes of these projects. In no case can it be clearly demonstrated that decisions would have gone differently in the absence of their efforts.

However, the advocacy movement has had a significant role in shaping the space community's culture, albeit gradually over more than three decades. This has occurred through a multitude of conferences, workshops, publications, and other formal and informal contacts across programs, disciplines, national borders, and generations. The effect undoubtedly has been far more subtle and more sluggish than some groups have intended, but it is recognizable and is moving in the direction they intend. The proof is in today's cadre of astropreneurs, the majority of whom came of age in an environment heavily influenced by the space advocacy movement.

Space tourism is perhaps the best example of an idea that needed to be cultivated in the outsider community of advocates before being accepted in public policy. Decades of evolution in technologies, economics, and attitudes were needed to make space tourism an accepted activity that governments could no longer ignore. In the past few years, the U.S. government has recognized that this is an endeavor to be nurtured, requiring near-term attention on licensing and regulation. Thanks to the efforts of enthusiasts and renegades, space tourism seems to be following the rule of thumb that says an innovative idea will become reality about 10 years after it overcomes the giggle factor.

A space business sector that is energetic, entrepreneurial, and technically and financially strong will be essential to ongoing operations that produce economic value and propel continuing development in space. Private sector investment and participation will be essential to accomplish and sustain activities such as crew and cargo transport, resource extraction, materials processing and manufacturing, off-world construction, energy production and distribution, on-site research and engineering, as well as terrestrial functions like mission planning, payload preparation, and personnel training. Without private sector help, the resources available from NASA—indeed, the combined resources of all of the world's civil space agencies—will be insufficient to ensure sustained off-world operations on the Moon and elsewhere in the solar system.

Unfortunately, there are some among the astropreneurs who shun the notion of partnering with, or even learning from, the government space sector. They view the government's half-century of experience in spaceflight with disdain, emphasizing cost overruns, performance failures, and dead astronauts while ignoring brilliant innovations and achievements. The vocal minority who hold this counterproductive attitude need to do some rethinking. Although their government counterparts are accountable to different stakeholders (taxpayers) and are motivated by a different mission (civil service and public safety), that doesn't make their expertise any less valuable. As noted in the previous chapter, government workers can be high achievers and innovators, just as private-sector workers can, if provided with adequate resources and the right incentives. Instead of rejecting their participation in the development of space, it would be wiser to cultivate partnerships with government organizations, and where possible, encourage a political environment in which they can prosper. In return, they'll do the same for the space business community.

The government responsibility to "encourage and facilitate"

Since the 1980s, it's been U.S. national policy to "encourage and facilitate" space commerce. You'll find this assertion in one or more presidential directives from each administration since that time, as well as in relevant laws and congressional reports. You'll find it in NASA's amended charter, which directs the agency to "seek to encourage, to the maximum extent possible, the fullest commercial use of space." However, policy-making on the commercial development of space has been at times inconsistent, ineffective, counterproductive, and partisan. Private sector interests often have had to deal with indecision, delay, and reversals in policies affecting their business plans—which explains why some astropreneurs have such a negative view of government "help."

In the first decade of the U.S. space program, NASA did not advocate the commercialization of space, nor was it directly concerned with economic development. Although technology transfer for the benefit of society was among the space agency's original mandates, NASA was not called upon to demonstrate economic gains to justify its efforts until Apollo was winding down and planning for the space shuttle was beginning. But proclamations in policies and laws since that time didn't automatically usher in a cultural change that would turn NASA into a world-class business incubator. The agency was already trying to do too much with too little, so the resources, incentives, and detailed guidance to make this happen just weren't there.

Is it even possible for a government agency to "encourage and facilitate" an emerging industry, or is "do no harm" the best that can be hoped for? Some who lament the slow pace of space development have suggested that the government end its space efforts (other than basic science) and let the more efficient, market-driven private sector take over. I believe this would slow the pace even further.

When discussing the unsatisfactory rate of space development, a favorite tactic is to compare it to the aviation industry. The time between the Wright Brothers' first flight and the first trans-Pacific passenger flights on Pan American Airways was a mere 32 years. It's been more than 40 years since the first humans walked on the Moon, yet we have no routine passenger flights to the Moon, or even to Earth orbit. "Unacceptable!" cry the critics. "This is clearly the fault of [NASA, the government, *insert your favorite scapegoat*]."

Such arguments are unsupportable, since space operations cannot be expected to follow development schedules similar to unrelated and technically less demanding activities. Aviation and space can't be directly compared just because both involve flight. (Apples and oranges are both fruits, but we don't compare them directly.) Indeed, even the various activities within the category of space commerce—launch services, communications, remote sensing, navigation—evolve technologically and operationally at vastly different rates.

In the early 20th century, the emergence of commercial aviation required the establishment of runways and airport facilities at all the destinations to be served. This was a formidable challenge in infrastructure building, but it could have been worse—the industry was not required to create the destinations themselves. These already existed, scattered generously all over the globe, complete with people, buildings, food, air, and other amenities essential to human activities. It's a different story in space, where we must escape Earth's gravity to travel, live, and work in an environment that is starkly different and more forbidding than any we experience on our home planet. Destinations like space stations and lunar bases need to be built, and critical supplies like food and air are not available locally. As a colleague of mine is fond of saying, space is always trying to kill you and your spacecraft. This makes it a bit tougher to devise and execute a viable business plan.

Some who accept the difference in the technical degree of difficulty between aviation and space still maintain that sluggish space development is the result of the dominance of the government, while aviation progressed faster because it was developed by private industry. This isn't just an oversimplification—it's inaccurate.

Aviation had its roots in entrepreneurial ventures and privately funded prizes for the setting of flight records, such as the Orteig Prize that inspired Charles Lindbergh to make his 1927 solo flight across the Atlantic. But it also benefited from government-subsidized activities such as military aircraft development and the research infrastructure provided by the National Advisory Committee for Aeronautics, the predecessor of NASA, beginning in 1915. Additionally, in the early days of the airlines, the U.S. government aided the struggling new industry by purchasing air mail services even though initially they were less efficient and less reliable than traditional means. Clearly, the success of the aviation industry is attributable to a cooperative mix of government and private efforts.

The same is true for space development. The government influence in this area is self-evident, but the private-sector contributions are also prominent in activities such as satellite communications, launch services, remote sensing, and global positioning services. Commercial efforts have profoundly affected the development of space since the early days, and will continue to do so at an increasing rate.

What should we learn from the history of aviation and space? Not that the two should follow similar development schedules, which obviously is not the case. Rather, we should observe that success lies in finding the proper blend of efforts between the government and non-government sectors. The path to successful space exploration and development is not a black-and-white choice between government agencies or private companies, nor should one or the other be blamed because big profits from asteroid mining and luxury condos on space colonies have yet to appear. The suggestion from some analysts that a government exodus from space development would open the floodgates of private investment is completely unrealistic, and will be for many years to come. The private sector has shown no indication as yet that it is willing and able to independently fund and operate all the spaceports, launchers, tracking systems, space platforms, and research labs necessary to maintain the current level of activity and drive expansive dreams of space commerce. Space development is more difficult, more time-consuming, more costly, and potentially more important than anything we have for comparison. Our task today is to build the programs that will systematically shape the future in space using the best efforts of all sectors of society.

After the Bush administration announced its space exploration policy in January 2004, it established an ad hoc advisory group called the President's Commission on Implementation of United States Space Exploration Policy (known as the Aldridge Commission after its chairman, former Air Force Secretary and Defense Undersecretary Pete Aldridge). In my estimation, the Commission's most important finding in its June 2004 report was that "NASA's relationship to the private sector, its organizational structure, business culture, and management processes—all largely inherited from the Apollo era—must be decisively transformed." The Commission recommended that "NASA recognize and implement a far larger presence of private industry in space operations with the specific goal of allowing private industry to assume the primary role of providing services to NASA, and most immediately in accessing low-Earth orbit." It also recommended that NASA set up "a technical advisory board that would give the Administrator and NASA leadership independent and responsive advice on technology and risk mitigation plans" and "a research and technology organization that sponsors high risk/high payoff technology advancement while tolerating periodic failures." Other noteworthy comments related to NASA-industry relations include the following (emphasis added):

> The Commission believes that *commercialization of space should become a primary focus* of the vision, and that the *creation of a space-based industry* will be one of the principal benefits of this journey. One of the challenges we face is to find *commercial rewards and incentives in space*. Creating these rewards is an indispensable part of making this partnership work in the right way. It will signal a major change in the way NASA deals with the private sector, and the Commission believes that NASA should do all it can to *create, nurture, and sustain* this new industry . . .
>
> What is impossible today will, in time, be commonplace. What is inherently governmental today will also change with time . . .
>
> We cannot overemphasize the impact a transformed NASA will have on the economy at large, if NASA truly commits itself to this new relationship . . .
>
> NASA must begin not only to *utilize private sector launch enterprises* more systematically, its exploration architecture must *systematically support private sector capabilities* that will make it possible to sustain operations in space . . .
>
> The Commission is convinced that *NASA's business culture must be changed* to embrace a significantly different role for itself in our space exploration enterprise. NASA needs a much-improved capability both to *learn from and partner with a more robust space industry*.

NASA has made some progress in this direction, primarily in its solicitation of private launch companies to service the International Space Station. However, the agency is still a long way from the institutional structure envisioned above, capable of routinely facilitating industry partnerships and willing to put privately owned and operated hardware and services in the critical path of its programs.

The Aldridge Commission didn't provide detailed guidance on how to execute these recommendations, nor was it asked to follow up on them at a later date. Nevertheless, the need remains, and it's not too late to implement these ideas. The suggested technical advisory board should consist of business-savvy professionals, but their duties don't have to be limited to advising the NASA Administrator. This external group could be closely linked with an internal office staffed by people with private-sector experience who would manage the agency's integration with industry. This internal office could be created by expanding the mandate of the existing Innovative Partnerships Program. In the future, NASA must be prepared to routinely make judgments that will be interpreted as endorsements of particular companies or technical paths serving space markets, such as who receives government assistance and who doesn't. This places NASA in the position of making industrial policy decisions on who gets to develop space infrastructure and resources. This is likely to raise objections from some members of Congress and from businesses whose proposals are rejected. Such situations are inevitable, and NASA will need to defend its decisions by demonstrating that they were arrived at through careful examination by a seasoned cadre of business professionals inside and outside the agency.

Of course, NASA isn't the only federal agency obliged to encourage and facilitate space commerce in accordance with national space policy. The private sector will need help from other parts of the government if it is to carry its weight and even take the lead in some areas. That will require additional incentives to prompt investors to put their capital at risk. This will involve the Congress and several federal agencies addressing a combination of legal, regulatory, procedural, and tax considerations, such as these that have been widely discussed in the space community:

- Reconsideration of export control restrictions on space products and services is needed to ensure that the international competitiveness of U.S. space industry is not being undermined by constraints that don't really serve national security. (See Chapter 6 for further discussion of this topic.)
- Commercial satellite operators have increasing needs for space tracking capabilities as orbital traffic and orbital debris continue

to grow. The world's most extensive space surveillance network is operated by the U.S. Defense Department, which has been providing limited support to commercial operators. But the system was designed to satisfy national security requirements. Some mechanism is needed to service the private sector without disrupting military functions, perhaps using an independent third party with the appropriate expertise to administer this technical service. The private sector doesn't want to duplicate the worldwide space tracking facilities that are already in place.
- An indemnification regime like the one maintained for the commercial launch industry to limit its third-party damage liability may be necessary for other types of space businesses. Liability risk will be a significant concern in corporate business plans, especially for activities which have little or no experience base and may not have commercial insurance available.
- Tax breaks on revenue produced from products made in space may help fledgling projects, if properly implemented. The tax savings should include carry-over into future years so that small businesses, which are less likely to have taxable income in their early years, would be able to benefit from the tax break.
- The issue of property rights in space has remained unresolved for decades. It's been pushed down the road each time it has come up because it's complicated, involves international negotiations, and has lacked urgency. But time is running out. This will take many years to resolve, and commercial entities will be reluctant to take on certain tasks—for example, resource extraction or assembly of infrastructure elements on the Moon—until this is settled.

A history of hype and hope

Just as government needs to do its part in shaping a business-friendly environment, astropreneurs need to do their part to create workable business plans, attract investors, develop products, and cultivate the market. In addition to learning lessons from government space experience, they should also heed lessons from the mistakes and miscalculations of earlier space entrepreneurs.

The 1980s was going to be the Golden Age of Space Commerce as a result of a confluence of circumstances: the initiation of space shuttle operations in 1981; the advent of the space station, which was expected to begin on-orbit construction in 1992; enabling legislation for new space ventures and privatization of existing government systems; substantial investments by a variety of business enterprises large and small; and burgeoning markets for space products. But aside from the continuing advances in satellite communications, space commerce fell far short of expectations. The space shuttle was not a boon to commercial users, and failed tragically in 1986; the space station experienced long delays and vastly reduced capabilities; enabling legislation for commercial launch services and satellite remote sensing was passed, but was so flawed that little happened until the laws were amended several years later; many ventures failed and investors lost confidence; and markets evolved very slowly or not at all.

The first U.S. commercial launches were ready to commence by the end of the 1980s, but two other hotly anticipated space markets—remote sensing and microgravity materials processing—left advocates coldly disappointed. In both cases, the technology wasn't mature enough, but the government didn't invest adequately in research, hoping to turn over this responsibility to the private sector. Also in both cases, the intended market was either uninformed about the new space product, or was skeptical due to concerns about quality, availability, and cost. This stands in sharp contrast to the communications market, which 20 years earlier had quickly welcomed its new space-based capability.

U.S. remote sensing commercialization was plagued by bad enabling legislation (passed in 1984, then repealed and replaced in 1992), dependence on government satellites (the Landsat series) that were never intended to be a high-volume operational system, and the inability of the government and industry partners to get along and fulfill their obligations. Things slowly improved after the passage of the 1992 legislation, helped by improved technologies that permitted private companies to develop their own purpose-built satellites. Today, the U.S. commercial remote sensing industry is beginning to live up to its promise, although about half of its revenue comes from bulk-purchase contracts from government customers.

Materials processing in space, or MPS as it was called, is a particularly interesting case because despite its great potential, it still has made no progress in developing commercial products. It remains at the basic research stage due to lack of funding and insufficient access to space-based laboratories and production facilities. Excessive hype by advocates in the 1980s raised expectations that were quickly dashed, damaging credibility with the investment community for a long time to come.

From the time MPS experiments were performed aboard the Skylab missions in the early 1970s, researchers in biological and materials science disciplines have been intrigued by the possibilities of using the unique environment of space to study and process materials. Among the applications touted are new metal alloys with special properties, purer pharmaceuticals, and better crystalline and ceramic structures. In addition to applied research leading to marketable products, basic research on the structure and behavior of matter in some cases can be done better in orbit, where the absence of gravity can reveal previously unknown characteristics that will allow advances in Earth-based applications.

The advent of the space shuttle raised hopes that large-scale production of profitable materials was just a few years away. The shuttle would serve as a space laboratory with materials science experiments as frequent flyers, ushering in a new sector of profitable space commerce by the end of the 1980s. Unfortunately, the shuttle didn't fly as often as advertised, and only occasionally carried a full laboratory, so many experiments were limited to the space available in the middeck lockers. Even these experiments were expensive to prepare and took a long time to get onto the cargo manifest. Many developers gave up on space research and opted for cheaper, more flexible terrestrial alternatives that would allow them to repeat experiments as needed rather than waiting years for a flight opportunity. The hope of near-term profits faded, no marketable products resulted, and the community still awaits better opportunities for performing research and eventually scaling up to production levels.

Given the cost of access to space, any viable product would have to be high in value and low in volume, as well as being economically competitive with, and/or technically superior to, Earth-based alternatives. Pharmaceuticals were early candidates to fulfill these requirements, including drugs to treat or cure common ailments like diabetes. One of the most celebrated projects was a collaboration between aerospace company McDonnell Douglas and the pharmaceuticals division of Johnson & Johnson. Together, they signed a Joint Endeavor Agreement with NASA in 1980 to pursue a microgravity purification process called continuous flow electrophoresis. McDonnell Douglas employee Charlie Walker flew as a payload specialist aboard three space shuttle flights in 1984-85 to conduct experiments with a $700,000 apparatus mounted in a middeck locker. The company also developed a processing device reportedly costing $11 million and large enough to take up 60 percent of the shuttle's cargo bay.

Although this collaboration was the most visible at the time, it was far from the only one. Numerous other companies flew research payloads, including recognizable names such as DuPont, 3M Company, Eli Lilly, Upjohn, Burroughs-Wellcome, and Kodak. The Center for Space Policy (now called CSP Associates), a consultancy in Cambridge, Massachusetts, forecast in 1984 that by 2000 the annual revenues for materials processing in space would be $41.6 billion, including $27 billion from pharmaceuticals alone.

These ambitions were unraveling even before the January 1986 space shuttle *Challenger* accident. Johnson & Johnson dropped out of its partnership in 1985, having decided that ground-based approaches were a better investment. McDonnell Douglas, unable to find a new pharmaceutical partner, called it quits soon afterward, selling its MPS flight hardware to NASA for a token payment and taking a tax write-off. Other companies were discouraged after *Challenger* by less frequent flight opportunities and more stringent conditions for getting their payloads onto the shuttle.

MPS was one of the original justifications for the space station, and was endorsed by the national space policies of the Reagan and Bush (Sr.) administrations. NASA promoted the concept to industry for 20 years, but then suddenly withdrew its support before any revenues were generated when the George W. Bush administration issued its U.S. Space Exploration Policy in January 2004, which directed that the space station will focus on "supporting space exploration goals, with emphasis on understanding how the space environment affects astronaut health and capabilities and developing countermeasures." MPS had fallen off NASA's radar screen. The National Institutes of Health has partnered with NASA to pursue interests in space biomedical research that may bring benefits to Earth, but the modestly funded program has moved slowly.

The combination of marketing hype, insufficient access to research facilities in space, and inadequate government investment has forced private-sector MPS efforts to wait for a time when their star would rise again and long-duration flight experiment opportunities would emerge. Hopefully, today's astropreneurs have learned from this and will avoid similar pitfalls. But the temptation to engage in marketing hype is always looming as competitors make bold claims about cost, schedule, performance, and market potential. Nowhere is this more evident than in space tourism.

Many developers of next-generation space launch capabilities are targeting, even counting on, the market for thrill rides into space for paying customers. If human-rated space launches can be made safe, reliable, frequent, comfortable, and affordable, then a large market could develop for passenger trips to orbit. Initially, significant numbers of people could make suborbital flights (ascending to an altitude above 100 km and immediately coming back down) just for the experience of space travel and the chance to look back at the Earth from space. Eventually, one can imagine a competitive market for off-world resorts in orbit or on the lunar surface.

Several individuals have purchased rides to the International Space Station aboard Russian Soyuz capsules. A great many others have put down deposits for suborbital flights, particularly with Sir Richard Branson's Virgin Galactic company, which was reporting about 300 reservations in mid-2009. Some enthusiasts interpret this as a demonstration that space tourism has arrived, but it's more accurate to say that it has arrived for people wealthy enough to spend anywhere from tens of thousands to tens of millions of dollars on adventure travel. That's per person, round trip, and does not necessarily include in-flight meals or the extra charge for your first piece of luggage.

Market surveys have found that large numbers of people would relish the awe-inspiring experience of looking down at the Earth and up at the heavens from orbit, while playfully engaging in microgravity acrobatics. Some industry pundits have been claiming that there will be hundreds or even thousands of space tourists (or spaceflight participants, as they are more formally called) flying each year by sometime in the decade after 2010. This is dependent on a lot of technical, regulatory, and market factors, many of them beyond the control of the spaceflight companies. Safety is the paramount concern, so one can't help but ask what the consequences would be if a spaceliner had a fatal accident. Would the market dry up overnight? Would lawsuits devastate the launch companies and spaceports, despite the fact that participants are required to sign liability waivers? Would the FAA Office of Commercial Space Transportation, which has regulatory authority, be compelled to suspend activities industry-wide, pending an investigation? How long would it take to recover, and would investors be patient?

For flights to orbit or beyond, there's also the question of what happens to public enthusiasm when people become fully aware of the less glamorous aspects of being an astrotourist: a better than 50-percent chance of getting space sick; the risk of radiation exposure, especially beyond low Earth orbit; the muscle atrophy and bone calcium loss that results from extended exposure to microgravity; the constant noise from ventilation systems and other equipment; cuisine worse than a high school cafeteria; no full-body cleansing facilities; and toilets that are sure to be unpopular.

Let's think positively and say that the industry is skillful and lucky enough to avoid serious mishaps. When will the suborbital spaceflight market be saturated? Few people will take this trip more than once, unless prices drop considerably. Astropreneurs seem confident that the suborbital business can flourish until the industry is ready to offer orbital flights, but that's not simply an incremental step that can be achieved on a predictable schedule. Orbital spacecraft will be more complicated, their launchers will have to deliver several times more energy than suborbital vehicles in order to reach orbital velocity, and there will need to be an appealing destination once reaching orbit. The investment for all of this is likely to be more than an order of magnitude greater than that of suborbital launch services.

A successful space tourism industry offering journeys to Earth orbit or beyond should have the following characteristics:

- Commercial spaceflight providers would not be completely dependent on vehicles and facilities owned and operated by a national government, although a state or regional model resembling airports could be advantageous.
- Trips to orbit would occur on a more-or-less routine schedule. They would not be one-of-a-kind events that generate widespread media attention.
- Pre-flight preparation for passengers would require little training (not more than a day or two).
- Age, health, and weight restrictions on passengers would not be so strict as to deny eligibility to a significant percentage of the potential market.
- A variety of tour packages would be offered. These would include special deals for couples and groups, and would feature alternative destinations.

The complexity and the importance of offering a selection of tour packages should not be underestimated. Such options are expected of every segment of the tourism industry, and space will be no different. What *will* be different is the degree of difficulty in setting up such packages. The only destinations available for the foreseeable future are the International Space Station (small capacity, not very comfortable, extremely difficult to book reservations) and a continuous loop around the planet.

Personally, I have no interest in a quick up-and-down ride into space, which seems more like barnstorming than tourism. (I'm sure my friends in the industry are hoping that few people share my attitude.) Here's my idea of a great tour package: We take a shuttle from Earth to meet up with a much larger space cruise ship parked in orbit. The cruiser spends a day in Earth orbit so we can enjoy the spectacular views of the home planet and get acclimated to our new environment. Next, we break away from orbit for our three-day trip to the Moon, where we park in lunar orbit for a day of equally spectacular, extreme close-up views of the celestial neighbor we've watched in the night sky all our lives. Our arrival is scheduled for full moon phase, so the familiar Earth-facing lunar landmarks are in sunlight. After that, we make the three-day trip back to Earth, having collected our high-definition souvenir photos and holographic snow-globes depicting the Apollo 11 landing site. This may sound ambitious, but I expect the toughest part will be convincing my wife to come along. She gets nervous about air travel, and refuses to ride on helicopters because she thinks they're too dangerous. In the potential space tourism market, how many people share her apprehension?

I sincerely hope the space tourism industry is able to operate safely, achieve its business goals, and then set even higher goals. But I remain cognizant of the many technical and economic challenges that could quickly sink their plans. I hope they don't fall prey to hype, either of their own doing or created by media coverage, because it could earn them the same fate as microgravity materials processing in the 1980s: rapid demise as a business prospect and decades of delay in resurrecting the dream.

Space tourism and MPS are not the only emerging space capabilities that have prompted careful scrutiny of hype vs. hope. Perhaps the most promising space application that has been studied seriously throughout most of the space age, but has yet to reach fruition, is the solar power satellite (SPS), which also is referred to as space (or space-based) solar power (abbreviated SSP or SBSP). The concept, which involves a constellation of large spacecraft dedicated to collecting solar energy for use on Earth, has been around since 1968 when it was proposed by Peter Glaser of the engineering consultancy Arthur D. Little Company.

Solar arrays have powered spacecraft ranging in size from a couple of kilograms to the approximately 400,000-kilogram International Space Station, which produces over 100 kilowatts of power. Space age requirements have given impetus to efforts to make the solar technology smaller, lighter, and more efficient. Extensive study of the SPS concept, which envisions beaming power to Earth using microwaves or lasers, has been done since the 1970s by numerous government and non-government entities in many countries.

The technical characteristics and system architecture options won't be discussed here because they've been amply covered in numerous publications over the years, and because the focus here is the broader societal implications of such a development. The argument over SPS has become an almost religious battle between those who believe it's the smartest, most comprehensive energy solution available, and those who say it's an insanely expensive scheme that will never work. I favor the advocacy side of this argument, although I don't think SPS will be a quick, easy, or comprehensive solution. Suffice it to say that: 1) I like the idea of designing an energy production system around the large, natural fusion furnace at the center of the solar system, which is expected to last for a few billion years; and 2) I'm prepared to take the word of the learned engineers and physicists who have spent decades developing the theory and building the subscale demonstrations that show SPS can work, and that it won't fry birds and airplanes passing beneath it, or set fire to cities.

There is no reason to suspect that SPS is technically infeasible, since collection of solar power in space and transmission of power across distances are both demonstrated technologies. The questions that need to be answered have to do with whether the system can be scaled up enough and whether a transmission distance equivalent to one-tenth of the way to the Moon is practical. Before jumping to negative conclusions, doubters should recall that the history of the past two centuries is littered with respected authorities declaring the impossibility of things that became commonplace within a few years (such as heavier-than-air flight; computers that are small, powerful, and ubiquitous; and human spaceflight).

SPS opponents correctly point out that the enterprise will be hugely expensive and take a long time—which is true of all infrastructure developments of any consequence. So the real challenge is economics. The business case clearly hasn't matured yet, but to say that it never will is short-term thinking. We can't approach a problem by saying, "The technology won't be available for 40 years, so there's no point in investing in it now." That's a self-fulfilling prophecy, guaranteeing that the capability won't be achieved in a reasonable time.

The SPS concept has long been hampered by an unfavorable reputation based on the late-1970s baseline model studied by NASA and the Department of Energy and skeptically reviewed by a National Research Council panel. At that time, SPS was intensely fought by oil companies (which saw it as competition), environmental groups (which became convinced the power beams would harm wildlife), and competitors for public funding priority. Since then, new concepts have been proposed that entail automated assembly with little or no astronaut labor, and power conversion efficiencies as much as three times higher than the 13.5 percent that was assumed in the 1970s studies. Additionally, today's design concepts are not locked into the notion of deploying dozens of satellites the size of Manhattan with the goal of satisfying all electricity demand in the United States. Hopefully, some lessons were learned from the 1970s experience: selling a big, new, expensive idea as a comprehensive solution to all our energy needs, while underestimating the speed and cost of its deployment and the vehemence of its detractors, is a prescription for failure.

From an economist's perspective, SPS doesn't appear worthy of endorsement until it can demonstrate delivery of power to a terrestrial electricity grid yielding a consumer price of about 10 cents per kilowatt hour. Given the immense up-front investment and long lead-time, that price point is nowhere in sight today. SPS needs help with risk reduction and demonstration, just as satellite communications did decades ago. And as everyone in the space business knows all too well, the biggest economic roadblock to this and many other space development ideas is the cost of getting payloads into orbit, which hasn't gone down significantly since the beginning of the space age. Some believe that the hoped-for space business renaissance and projects like SPS will be realized only if launch costs are reduced by a factor of 10 to 100, and that this will not be possible using any means we know of today. I look at it this way: if we intend to be spacefarers and "incorporate the solar system in our economic sphere" as former presidential science advisor John Marburger has suggested, we'll continue our efforts to make access to space affordable, reliable, frequent, and flexible. If we give up on those efforts, it's a signal that we're giving up on space applications beyond those that have already become routine. I vote for continuing to move forward.

The electricity price point of 10 cents per kilowatt hour is a convenient metric, but does it make sense in this case? It would if U.S. electric utility consumers were the only target market, because that's near the price they're accustomed to paying. However, it isn't relevant to all parts of the world or all types of applications. And if energy costs climb, it won't be relevant to U.S. consumers anymore.

Let's look at SPS in a different way, as an energy utility with characteristics similar to another space-based utility: satellite communications, the poster child of successful space commerce. Like the power transmissions from SPS, communications are beamed down to Earth using microwaves, although the satellites are a tiny fraction of the size of proposed SPS platforms.

In the 1950s, satellite communications concepts were in their formative stages. There were a variety of notions on what a system design should look like, and what kind of organization should run it. Many technical aspects still needed work, and the boosters required for deployment in geostationary orbit were yet to be proven. There was no definitive evidence that satellites would ever be cost competitive with the large number of well-established ground-based options. But within a few years, a global industry emerged that today boasts remarkable success and continuing growth. This was able to occur because the product:

- consists of electromagnetic signals, not physical cargo, traveling back and forth from space;
- has near-universal demand, used by customers whether or not they're making money with it;
- entered a market with a well-established terrestrial infrastructure, both physical and regulatory;
- allowed developing countries to technologically "leapfrog" the costly and time-consuming intermediate development paths taken by industrialized countries.

All of these characteristics also can be attributed to space-based energy services. The SPS concept today may be at a stage similar to satellite communications in the 1950s: not ready for prime time, but demonstrating the potential to be an important addition to global infrastructure that addresses clear and present needs. However, one big advantage enjoyed by satellite communications, which space-based energy has yet to experience, was the visible support and substantial funding provided by the U.S. government at its inception. SPS needs government help in scaling up and demonstrating the technology. Ultimately, each platform would need to provide something in the range of 1000 to 10,000 times the output of the International Space Station power system, which is based on 1980s technology.

Recent projections from the U.S. Department of Energy foresee world electricity demand growing 77 percent between 2006 and 2030. Developing countries will exhibit almost three times the annual growth rate of industrialized countries during this period. This remarkably swift demand growth, coupled with the environmental and climate concerns that accompany expansion of the energy sector, compel us to look beyond expedient short-term solutions. SPS, like satellite communications, navigation, and remote sensing before it, could provide a new space-based capability that is important for industrialized nations, which are being asked to curtail emissions of carbon dioxide and other greenhouse gases, and for developing nations, which may be able to employ SPS to leapfrog the energy generation technologies that are responsible for a significant portion of harmful emissions and other types of environmental decay. The SPS concept is worthy of greater attention from both the public and private sectors, as will be discussed further in Chapter 8.

Vision or fantasy?

Elaborating on the question posed in the title of this chapter: Is development through commercial activity the real long-term vision for space, or is it a fantasy that will play out mostly in briefing charts and promotional videos, never really achieving sustainable value generation?

Many of today's space visionaries are found in the commercial sector, judging from all the attention in recent years to suborbital space tourism, privately-owned orbital habitats, prize competitions, online map services using satellite imagery, and new applications in mobile devices that depend on communications and navigation satellites. But the commercial space sector is no stranger to overpromising. As we saw in the 1980s, there were many assurances of substantial near-term profits in launch services, remote sensing, and microgravity materials processing that were not realized, driving investors elsewhere. Commercial space visionaries should not be too sanguine about their ability to turn their dreams into realities more effectively than their counterparts in the government arena. Both sectors face the difficulty of promoting very long-term visions in an instant gratification society. Both need the creative thinking and multidisciplinary approach that constructive partnerships can offer.

The private sector has the technical capability, and given proper motivation and some help with risk mitigation, can aggregate substantial resources for efforts of very large scope, as noted in Chapter 1 with the examples of the Alaska Pipeline and the English Channel Tunnel. However, the timing and amount of returns that can be expected from most space investments are less certain than those that have resulted from the Alaska Pipeline. Corporate interests cannot be confident that they will get back what they've invested within a reasonable timeframe.

The question posed by this chapter can't be answered yet. Instead, it prompts more questions. Can the public and private sectors formulate a lasting partnership that maintains a long-term view? Can multiple sustainable space-based applications be developed that yield attractive returns? Will disruptive technologies appear that profoundly affect space commerce, for better or worse? Can milestones along a clear path to strategic and economic objectives be established, yet remain flexible enough to accommodate a variety of probable long-term scenarios?

Chapter 6
Be Careful What You Wish For

Every one desires to live long, but no one would be old.
—Abraham Lincoln

As noted earlier, futurism isn't about predicting what will happen by a particular date. While it may be interesting and fun to do so, the real purpose of futurism is to shape the future. We've seen that this is best achieved in an environment that encourages creativity and where the barriers and disincentives to forward thinking have been minimized. We have a lot of work to do before we achieve that environment, and once we do, we're not finished yet.

In a futurist-friendly world, investigating major trends and extrapolating the multitude of possible paths that could result will be a more common and more sophisticated activity than it is today. The task for planners and decision-makers will be to determine desired outcomes that may take decades to come to fruition, then establish the required milestones along the path from here to there. Ideally, the planning process should thoroughly examine the possible consequences of the goal so that time and resources are not wasted achieving something that is not useful, or may even be detrimental, when it finally appears. Conceptually, this seems simple and obvious, but history has shown that making wise choices can prove elusive, yielding disappointing and costly results even in cases where a lengthy and seemingly thorough analytical process had been conducted.

Lessons learned (hopefully)

History is riddled with stories of unintended consequences, many of them related to explorers and colonists moving into unfamiliar environments and wreaking havoc. Early European settlers in the Americas brought with them diseases that devastated native populations. Immigrants to foreign lands have often tried to behave just as they did back home, bringing flora, fauna, and traditional farming techniques to places with completely different climates and ecosystems. One of the stories Jared Diamond tells in his book *Collapse* is about English settlers in Australia who tried to do this starting in the late 18th century. Among the many undesirable results was an uncontrollable proliferation of rabbits and foxes that has been causing ecological and economic damage ever since.

In the days when we still used to talk about the "conquest of space" there were times when we learned the hard way about the implications of certain behaviors in the space environment. A test of a 1.4 megaton hydrogen bomb above the atmosphere in July 1962, called Starfish Prime, disabled several satellites and taught us that electromagnetic pulse radiation could interfere with electrical systems across a wide area on the Earth below. The test was conducted even though U.S. and British scientists, whose input was not considered important by decision-makers, predicted the unwelcome effects. Around the same time, anti-satellite projects (Bold Orion, Nike-Zeus, and others starting in 1959) and a bizarre experiment attempting to relay communications signals by bouncing them off an orbiting cloud of millions of needles (Project West Ford, May 1963) gave us our first lessons on the dangers of orbital debris. Specifically, they showed us that manmade debris is more hazardous than natural micrometeorites, which just pass briefly through the vicinity of Earth rather than accumulating in orbit.

Human spaceflight has gone through high-cost, high-profile trial-and-error learning experiences as well. The decision to build the Space Transportation System—the space shuttle, as it's more commonly known—came in January 1972 after long consideration of its technology and economics. Technically, the idea of a reusable spacecraft that could land like an airplane had been around since before the start of the space age. Economically, it was seen as a way of bringing down launch costs and increasing the flight rate. Politically, the debate on what kind of human spaceflight project should follow Project Apollo, including its goals and how much should be spent on it, had been going on for a few years as well. After weighing the merits of various NASA design concepts, DoD's requirements, and White House budget plans, the choice was made on the launch vehicle that would be the basis for U.S. space efforts for decades to come. The development cycle turned out to be longer and more difficult than anticipated. But that's to be expected in cutting-edge developments, and we can learn a lot from it. After all, President Kennedy said we do these things "not because they are easy, but because they are hard." When you're building systems that provide critical infrastructure and need to last for decades, it's more important to get it right than to do it fast—in this case, to build the right vehicle for the right reasons.

Fast-forward three decades. Just about everybody in the space community at the turn of the new millennium agreed that the shuttle should be retired soon and replaced with something better that would be cheaper to operate. Disturbingly, many people in the community—including the NASA Administrator in 2005—went even further, saying that the space shuttle, and the space station program that followed it in the 1980s, were mistakes because they confined our human spaceflight missions to low Earth orbit. If we assume that this assessment is correct, it casts a negative light on the long-term planning process that initiated these programs. We might ask, what were they thinking back then? That's actually a long story, mostly concerned with post-Apollo attempts to ensure the continuity of human spaceflight, and indeed, the continuing justification for a civilian space agency. It had very little to do with a long-term strategic plan for human expansion into the solar system. An underfunded shuttle development effort became NASA's flagship program, with the goals of bringing down the cost of access to orbit, increasing flight safety, and conducting dozens of flights per year. Its debut in 1981 brought impressive new capabilities, supporting a crew of up to seven astronauts and featuring a cargo bay big enough to hold a school bus—a huge leap from the Apollo capsules of a few years earlier that offered seating for only three. But despite being an engineering marvel, it never achieved its promised goals.

A more recent example of questionable strategic planning for human spaceflight occurred after Mike Griffin took over as NASA Administrator in April 2005. He immediately began a planning effort to develop a new space transportation architecture for returning to the Moon, and rolled out the new architecture just five months later. Then in April 2006, seven months *after* the unveiling of the transportation architecture and more than two years *after* the announcement of the lunar exploration program, NASA put out a solicitation seeking ideas on what to do when we get to the Moon. This is backwards logic, like building a trucking fleet for a company before you know what line of business the company will pursue.

The repeated reassessments and reconfigurations of the International Space Station (ISS) program provide an ideal example of what can happen when long-term strategic planning takes a back seat to short-term political and programmatic goals. The stated purpose of the space station has changed numerous times since the program took shape in the early years of the Ronald Reagan administration. Before approving the program, President Reagan ordered an interagency study "to establish the basis for an Administration decision on whether or not to proceed with the NASA development of a permanently based, manned Space Station." Foremost among the policy issues to be studied were the station's contribution "to the maintenance of U.S. space leadership and to the other goals contained in our National Space Policy" and how the station would "best fulfill national and international requirements versus other means of satisfying them." An opportunity to layout a relevant, broadly accepted, long-term plan for the program was missed when the interagency working group failed to reach consensus and the study was never completed.

The administration's space station concept began as a multipurpose facility motivated by a desire to match or exceed Soviet capabilities in human spaceflight. It was going to be all things to all people. According to NASA public affairs literature in the mid-1980s, it would be:

- A national laboratory in space, for the conduct of science as well as the development of new technologies and related commercial products;
- A permanent observatory, to look down upon the Earth and out into the universe;
- A servicing facility where payloads and spacecraft are resupplied, maintained, upgraded, and if necessary, repaired;
- A transportation node where payloads and vehicles are stationed, processed, and propelled to their destinations;
- An assembly facility where, due to ample time on orbit and the presence of appropriate equipment and personnel, large structures are put together and checked out;
- A manufacturing facility where human resourcefulness and the servicing capability of the station combine to enhance commercial opportunities in space;
- A storage depot where payloads and parts are kept on orbit for subsequent deployment; and
- A staging base for future endeavors in space.

In addition to these impressive technical capabilities, the station would serve political and diplomatic purposes as well. It would be "a striking example of Free World unity and capabilities" and "a highly visible yet peaceful demonstration of U.S. leadership."

These ambitious goals did not even survive through the end of the Reagan administration. By 1987, tight budgets and a review by the National Research Council drove the program to its first major redesign. The so-called "revised baseline" cut back on accommodations for attached payloads for Earth and space observations and eliminated capabilities for satellite servicing, construction of large space structures, and staging of manned missions to the Moon or the planets. In other words, the eight major functions touted by NASA were reduced to just one: a national laboratory in space.

Lower expectations for the space station program have been reflected in the policies of successive presidential administrations. The national space policy issued in the last year of the Reagan administration, when the station was still in its early developmental phase, stated the objective of establishing "a permanently manned presence in space." The station was expected to:

- Contribute to U.S. preeminence in critical aspects of manned spaceflight;
- Provide support and stability to scientific and technological investigations;
- Provide early benefits, particularly in the materials and life sciences;
- Promote private sector experimentation preparatory to independent commercial activity;
- Allow evolution in keeping with the needs of station users and the long-term goals of the U.S.;
- Provide opportunities for commercial sector participation; and
- Contribute to the longer term goal of expanding human presence and activity beyond Earth orbit into the solar system.

This list was less specific and more limited in scope than NASA's initial vision for the program, but still featured a formidable array of functions. The space policy of the George H. W. Bush administration, which appeared in November 1989, retained the same wording as the Reagan policy relevant to the space station.

President Bill Clinton's national space policy of September 1996 continued the process of limiting the expectations of the space station program. Gone was any specific mention of benefits from materials and life sciences research or endorsement of private sector participation. The only discussion of the purpose of the space station in the Clinton policy directs NASA to:

> Develop and operate the International Space Station to support activities requiring the unique attributes of humans in space and establish a permanent human presence in Earth orbit. The International Space Station will support future decisions on the feasibility and desirability of conducting further human exploration activities.

Although not mentioned in his national space policy, President Clinton clearly had foreign policy goals in mind for the space station. During his first year in office, he invited Russia to join the other international partners in the station (Canada, Japan, and the European Space Agency) to improve relations and keep Russian expertise from being diverted to entities unfriendly to the United States.

The refocusing of space station program goals by the George W. Bush administration endorses only the last item on the Reagan-era list: research toward "expanded human presence and activity beyond Earth orbit." Along with a commitment to complete the station and honor U.S. obligations to international partners, the January 2004 presidential directive on U.S. space exploration policy declared that U.S. research and use of the station would focus on "supporting space exploration goals, with emphasis on understanding how the space environment affects astronaut health and capabilities and developing countermeasures." This tied the ISS directly to the fate of the administration's exploration program. Even more troublesome was that the Bush administration left office with no plan to continue U.S. use of the station after 2015. Apparently, a quarter century of effort and investment were to be simply written off, and the keys to the facility handed over to the international partners.

While the stated purpose of the ISS steadily shrank across two decades and four presidential administrations, Congress didn't exercise leadership in halting the slide of the program's strategic goals or research objectives despite its control over the station's budget. Confronted by technical challenges, competing scientific priorities, and a changing political environment that included the end of Cold War competition, Congress focused on cost and schedule problems while leaving the task of defining the station's purpose to NASA and the administration.

From a strategic planning perspective, the ISS experience illustrates a failure to apply lessons learned from the space shuttle program. The shuttle also had tried to be all things to all people. It's understandable that numerous stakeholders would want to see their requirements accommodated in a new flagship project, and that program managers would see this as a way to expand the program's constituency. But this is not a practical approach when multiple conflicting operations are planned, such as conducting microgravity research, which must keep vibrations to a minimum, on the same space platform as life sciences research using treadmills and centrifuges. Inevitably, an all-in-one design approach will drive up complexity and cost, and may have serious consequences for the scheduling and quality of the research.

Alternative approaches were available to the planners and decision-makers of the early 1980s. One option for addressing the multiple-mission problem would have been to build a series of stations, perhaps based on shuttle external tank components, each one optimized for a set of related tasks. For example, there could have been a human-tended (that is, not continuously human-occupied) microgravity laboratory, a human-tended spacecraft assembly and repair station, and a permanently crewed variable gravity facility, which would spin at different rates to provide zero-g, lunar gravity, and Mars gravity environments for human physiology research. This would have allowed research and operational capabilities surpassing those of today's station to come online gradually, possibly reducing budget and schedule risk. Of course, this approach also would have given White House and congressional budget cutters three targets to shoot down one by one.

A quarter century of evolution in the rationale for the ISS, which was originally called NASA's "next logical step," illustrates that the initial program image of national pride, Cold War rivalry, and technological achievement gave way to diminishing goals that were a product of short-term thinking and insufficient analysis of options. After our experiences with Apollo, shuttle, and the space station, we have to ask ourselves how long we can continue repeating this pattern of experiments with multi-decade lifecycles and astronomical price tags. We develop a major infrastructure element, use is for 10 or 20 or 30 years, declare it a mistake, discontinue it, and start the process over. We've reached a stage where we can't afford this any longer, economically or politically, so we need to start doing a better job of thinking ahead—way ahead. It won't be easy.

Thought experiments

Let's engage in a brief futurist exercise to illustrate how tough it is to account for all of the significant implications of a major course of action. The two hypothetical issues that follow are not currently on the public policy agenda. They may never be—or they could become important issues sooner than we anticipate. The purpose here is to show the potentially disruptive effects of scientific and technological developments on society, and the challenge of formulating public policy and adapting institutions to cope with the changes. Following these two speculative cases, we'll look at a real world case of a policy change that has caused an assortment of ongoing problems for U.S. space industry because it was implemented for short-term gain and was poorly thought out prior to its initiation.

Do we really want to live forever? Life extension has been a human desire for as long as there's been death. In the long history of our species, only in the modern age has there been some measure of success, and we can reasonably expect significantly more. In fact, some futurists today, such as Ray Kurzweil, are telling us that if we can hang in there for another 15 or 20 years, we will have the opportunity to live forever. Forever? How many people have really thought about the implications of this, both for the individual and society? This is a good thought experiment for long-term strategic planning because it illustrates a goal that may seem highly desirable at first glance, but reveals some serious problems upon closer examination.

I've always felt that we need at least 300 years to learn about life, the universe, and everything, and still have some time to apply what we've learned. Aside from putting off death, the benefits would include the ability to engage in multiple successive careers to explore our interests and use our knowledge; the opportunity to know our descendants for more than just a couple of generations; the delight (hopefully) of observing the evolution of the arts, technology, and space exploration; the joy of becoming proficient on several musical instruments; and the chance to read all those books on the shelf that we never seem to get around to. And think of all the vacation time you could accumulate—maybe that 24-week around-the-world cruise isn't out of the question after all.

But what would be the implications for society as a whole? Already, we have grave concerns about the rapid growth of the world's population. If people lived multiple times the current life expectancy, and birth rates didn't drop precipitously, we wouldn't be able to deal with the resource demands, the environmental degradation, and the overcrowding that would result unless we expanded into the solar system and took advantage of the raw materials and energy available there. (Hmm . . . we'll come back to this idea later.)

That's not the only concern. The whole employment picture would change. People couldn't retire at 65 or 70 and expect to collect Social Security and pensions for potentially hundreds of years. We're already running out of young working people to support the retirees, and this scenario would quickly drive the system to the breaking point. So people will have to work longer—a lot longer. Our assumption is that longevity is accompanied by good health, so a longer working life may be welcome. Stay in the career you love! Go back to school and retrain for a new career! Who wouldn't be happy with that? I'll tell you who: young people.

Imagine an employment environment where job openings are rare because older employees stay on for a long, long time. Even if someone fresh out of school manages to get a job, it will be at the bottom of a ladder that will take a very long time to climb. After all, they could be competing with co-workers whose experience exceeds theirs by several decades. This could cause a great deal of resentment, even rebellion, in the entry-level workforce, which is where people could be stuck until they're at least 50. Or they could be out of the workforce entirely. High unemployment among the under-50 youngsters is a prescription for civil unrest, and may engender diminishing trust and respect for the older generation (over 100). The solution would need to be driven by two things: society's constant need for fresh ideas from the younger generation, and a vastly expanded economy that could absorb all of these new workers, perhaps driven by successful economic development of the solar system. (Hmm . . . there's that idea again. Longevity would be a big help given the commuting times involved.)

There will be difficult equity issues as well. Let's say the longevity comes from a magic treatment that stops the aging process and eliminates the threat of disease. Is it a one-time treatment, or does it need to be administered regularly? Most likely it would be the latter. Will the treatment's cost be affordable by the general population, or only by the rich? That will depend on whether or not it requires a scarce resource or is prepared using an expensive process.

If the anti-aging treatment is expensive, the traditional animosity between the haves and have-nots, both domestically and internationally, would reach an entirely new level. The haves would control not just the majority of the wealth, but the fountain of youth as well! Would the much more numerous have-nots stand for this? Would they forcefully try to take the elixir of youth for themselves, or failing that, attempt to destroy it so that no one could benefit from it? Similarly, religious fanatics could seek to deny life extension to "unbelievers" or, alternatively, condemn it as an evil act that contradicts God's will.

Societal adaptation to extended human life seems to be a more difficult prospect than we might have expected. In addition to the need to completely revamp the way we do life insurance, pensions, and social services, the economic implications include the fact that in a world where everyone's healthy, the health care industry—one of today's biggest growth industries—would suffer massive unemployment. (People who have had bad experiences in their medical care may not have much compassion for this situation.) But it wouldn't disappear completely. People would still have medical needs because in addition to administering life-extension treatments there would still be injuries to treat.

Speaking of personal injury, let's consider individual adaptations to longer life. If you're immune to aging and disease, that means you can expect to be around for a very long time, but it's not immortality. Everyone will (or should) realize that they are cheating death for some finite amount of time, not escaping it completely. Everyone should also realize that since old age and disease have been banished, they will ultimately meet their demise through accident or foul play. No one will rejoice over the prospect of a violent death, which will only be avoided if you're lucky enough to succumb quietly in your sleep at the ripe old age of 425 (after which 17 generations of your family will fight over your will).

In such a world, people may fear premature death even more than they do now, making them even more risk-averse than they are today. If enough people feel this way, it will have a chilling effect on experimentation, exploration, and even routine travel. Forget about those forays into the solar system—space travel is going to be dangerous for a long time to come. On the other hand, long-lived people could get bolder as they get older. Having logged in several decades of life, they may look for new and more daring experiences without fear that their already full lives would be unduly shortened. As healthy individuals who are more concerned about avoiding boredom, they may become the primary customers of the adventure tourism industry, and the favored candidates for high-risk or unconventional jobs. If this were to happen, it would be a complete reversal of the present situation, in which younger people tend to take the risks while older folks are more inclined to avoid them.

This brief discussion has shown that extreme life extension—a development that initially seems beneficial to all and to have no downside—actually does present some serious challenges at a variety of levels. As bizarre as parts of this scenario may sound, it illustrates the need for plenty of serious thought before launching into something that will involve substantial investment and will affect many generations to come.

Next, let's try a more fanciful, but still familiar idea to illustrate the importance and the challenge of thinking long-term.

The longing for sci-fi tech. Anyone who has even a passing familiarity with the *Star Trek* saga knows about a futuristic technology that is routinely employed in the stories: transporters. Possibly the most famous bit of technical wizardry associated with the *Star Trek* universe, these devices provide near-instantaneous delivery of people and cargo to destinations near and far. (At a maximum, interplanetary, but not interstellar, distances. Even science fiction imposes limits sometimes.) Similar devices can be found elsewhere in sci-fi, and they can take a variety of forms. The *Stargate* of movie and television fame, unlike *Trek*'s transporters, requires devices at both the sending and receiving ends. The same is true for the so-called "displacement booths" in the stories of science fiction author Larry Niven.

The societal implications of this type of transport have been considered by many people over the years. Niven wrote a thought-provoking and amusing essay in 1969 titled "The Theory and Practice of Teleportation" in which he borrowed from science fiction writers who came before him and added a few twists of his own; I'll borrow from Niven and do the same.

As with life extension, teleportation seems like a winner all around, as long as you have no qualms about having every molecule in your body turned into an energy stream and reassembled elsewhere. (Apparently, this bothered *Trek*'s Dr. McCoy, but not enough to keep him from using the transporters.) If it's affordable and capable of handling a fairly large amount of mass, society could do away with cars, trucks, trains, airplanes, and boats, except for those used for leisure or entertainment purposes. If the range of the transporter is sufficient, some types of spacecraft could be eliminated. This would cause some industries, such as auto manufacturing and everything that supports it, to shrink to insignificance, but at the same time it would allow the material, energy, and labor resources applied to those industries to be used elsewhere; would eliminate the pollution caused by fossil fuel-powered vehicles; and would open up possibilities in commerce and travel that would more than make up for the loss. And think of the increase in productivity if the time spent on business travel and daily commutes shrank to essentially zero. If you live in an urban area, fantasize this: no traffic jams and no overcrowded public transportation—ever.

I'm sure we can all come up with plenty of useful or even revolutionary applications for this capability. So what's the bad news? It would be the ultimate security nightmare. Imagine what it would mean for securing homes, businesses, sensitive government installations, even whole nations. Evil-doers could appear and disappear rapidly, foiling attempts at detection, response, and pursuit. Or they wouldn't have to show up at all—just send a bomb.

The types of safeguards required would depend on whether or not teleportation required devices at both the departure and destination points. Characters in Niven's stories quickly learned that they shouldn't install a displacement booth inside their house, since that could result in the unannounced and unwelcome arrival of burglars in their living room. Presumably, such a transport system would have its own version of Caller ID or some other screening capability, but that wouldn't prevent scam artists from entering people's homes under false pretenses, just like they do by knocking on the front door today. It's probably better to install the booth in a separate structure out in the back yard.

Clearly, the same concerns would affect businesses, government agencies, and any locations or events where large numbers of people gather. Transport booths would have to be placed somewhere other than the most convenient locations so the bad guys and their weapons wouldn't have it too easy.

The security problems become even tougher if you have a *Trek*-like transporter that needs a device only at one end of the trip. Your protection would have to come from another well-known *Trek* device, an energy shield (or alternatively, some kind of jammer). No location would be safe without it unless it was out of range or buried within enough matter that transport signals couldn't get in or out. It would be impractical for most locations frequented by people to be out or range or buried, so the shield becomes the only choice. So you'd need a shield for every home, school, shopping mall, sports arena, church, and office building. In fact, for office buildings and apartment buildings, you'd need a separate shield for every suite. What a mess. Sort of defeats the purpose of having the devices at all.

As with life extension, the social implications vary depending on whether the transport system is ubiquitous and affordable by all, or expensive and limited in its availability. If its use is as cheap and handy as cell phones, people could become inclined to live their whole lives without experiencing the outdoors, a possibility suggested by author Isaac Asimov. In that circumstance, would areas that are not served by the transport system become abandoned and forgotten, or would they be havens for vacationers and retirees who seek refuge from a hectic, globetrotting life?

Admittedly, the transporter scenario will only be confronted far in the future, if ever. But it provides an excellent example of the vast array of unintended consequences that can ensue from disruptive changes, and the difficulty in dealing with them at every level from the individual to the whole of society. Rather than being far-fetched, this is precisely the kind of experience we've been having since we began widespread use of the Internet.

Back to reality

The fanciful exercises above, though not intended to be comprehensive in scope, are enough to demonstrate that it's important to consider more than just immediate costs and effects. But we've already seen in the cases of the space shuttle and space station, and in countless other policy developments reported in the daily news, that this simple rule is often violated. Another clear example is one that goes to the heart of U.S. leadership in space technologies: export control.

During the late 1990s, when President Bill Clinton was facing an opposition Congress, Republicans sought to portray Clinton as soft on China. Their cause was helped by the February 1996 failure of a Chinese rocket launch carrying an American-made satellite. With pieces of sensitive U.S. technology splattered across the Chinese landscape near the launch site, fears of technology transfer that would aid development of China's weapon systems became more prevalent than ever on Capitol Hill. Soon after that incident, a U.S. satellite maker submitted technical information to a Chinese launch failure investigation but neglected to get an export license from the State Department for that information. As a result of these events, language was added to the fiscal year 1999 defense authorization bill to tighten restrictions on export of space technologies through the International Traffic in Arms Regulations (ITAR). In a throwback to the Cold War days, commercial communications satellites would be treated the same as weapon systems for export licensing.

Ironically, the Bush (Sr.) and Clinton administrations had just spent the better part of the previous decade planning and implementing a gradual move away from Cold War restrictions on technologies for which global markets were beginning to flourish. For space technologies, this movement culminated in the shift of export licensing control over commercial satellites from the State Department to the Commerce Department in March 1996. (Notably, this was the month *after* the infamous Chinese launch failure. That controversial incident occurred on the State Department's watch, yet the tighter export restrictions taking effect in 1999 would return licensing control to State.)

With the Soviet Union gone and globalization accelerating, the Bush and Clinton administrations recognized that although certain "crown jewel" technologies still needed to be protected, the United States was not a monopoly provider of most of the duel-use technologies on the evolving world market. ("Duel use" refers to technologies that are commonly used in applications in both the national security and the civil/commercial sectors, such as optical sensors, guidance systems, and data handling systems.) But the U.S. at the time did have an edge in many high-tech areas in quality, price, and ability to deliver quickly. In such an environment, these administrations recognized, the nation's security is better served by engaging in the world market in an area of strength rather than yielding all the business to increasingly capable foreign competitors, to the detriment of the U.S. industrial base.

Unfortunately, this logic was lost on proponents of tighter export control in the Congress. The immediate prospect of embarrassing the Clinton administration on this issue blinded opposition-party lawmakers to the ramifications of their actions, even though many analysts and industry representatives had warned of the drawbacks. Despite this ill-advised reversal of an export control reform process that had been years in the making, President Clinton did not use his veto power to prevent this action because he would have had to veto the entire defense authorization bill to do so. Also, if he had used his veto, the issue wouldn't have gone away. He would have been pilloried for both his policy toward China and his willingness to delay an important defense bill.

The legislation put commercial satellites, and most of the subsystems and components related to them, back under the jurisdiction of the State Department in March 1999 and tightened the controls on them, even when exporting to close allies. The effect on China was to delay by a few years its efforts to become a successful provider of commercial launch services, because all the established manufacturers of commercial satellites at the time were in the U.S. and Europe and used U.S. parts. The restrictions had no apparent effect on China's multifaceted space technology development efforts, which have demonstrated steady advancement since then. The consequences for the United States, on the other hand, were much more detrimental and lasting.

The prevailing view since 1999 has been that a problem was created rather than solved. Export control has evolved from a nuisance to a serious national problem. Media coverage of space technology export control issues since the stricter regulations went into effect has focused overwhelmingly on the difficulties and delays that have resulted. The trade press and the general media have featured headlines such as "Export Control Rules Strangling U.S.'s Future," "U.S. Satellite Sales Lag Since Regulatory Shift," and "European Satellite Component Maker Says It Is Dropping U.S. Components Because of ITAR."

Even without the regulatory changes, the U.S. was destined to lose global market share in space-related products because capable international competition was emerging even faster than the market was growing. But the timing and abruptness of the shift in market share for satellites, subsystems, and components was unmistakable: the new export rules were a game-changer. If declining sales figures weren't enough to prove the point, all the U.S. needed to do was listen to what foreign governments and companies were saying about phasing out U.S. components and avoiding partnerships with U.S. companies because of the added costs, delays, and hassles.

According to the Satellite Industry Association (SIA), U.S. satellite manufacturers' share of the global market for commercial satellites declined 30 percent in 2000, the year after the new export law took effect. That brought it down to just 45 percent compared to the previous 10-year average of 75 percent. While acknowledging that not all of the lost revenue and market share could be attributed to the new export controls, SIA found that the regulations were clearly having a negative impact on the ability of U.S. commercial satellite companies to compete.

Different studies show variations in the estimates of U.S. market share before and after the 1999 change in export controls. For example:

- A 2000 study by Booz Allen & Hamilton identified a "historical" market share of 70 percent.
- A 2002 study by the Center for Strategic & International Studies (CSIS) found 1997 to be the peak year for U.S. satellite manufacturing revenues, at 65 percent of the world total, declining to 50 percent by 2001.
- A 2006 article in the online journal *The Space Review* pegged the pre-1999 market share at 83 percent.

The differences are attributable to variations in the timeframes of the studies and their definitions of what is included in satellite manufacturing revenues. Nonetheless, all agreed with the SIA study that the regulatory change had significantly damaged the industry in the near term, and could continue to do so in the long term.

The Booz Allen study called for the executive branch and the Congress to form a consensus on the appropriate balance between technology protection and industrial competitiveness, and to revamp the license review process to distinguish between varying levels of technology sensitivity and end users. This recommendation foreshadowed a problem that later became evident: the movement of traditional U.S. customers to foreign satellite manufacturers, due at least in part to dissatisfaction with the added restrictions and potential delays. Examples include Telesat Canada, Arabsat, and Intelsat, all major satellite operators that switched from U.S. to European satellite manufacturers. As CSIS and others have pointed out, U.S. controls on satellite technology are significantly more restrictive than those faced by foreign competitors, resulting in a competitive playing field that is tilted against U.S. companies.

U.S. manufacturers became increasingly wary of bidding on certain foreign contracts. They passed up competitive bids because they anticipated some combination of export control problems that would significantly decrease their likelihood of winning the contract and/or their ability to perform on cost and on schedule if they did win. Manufacturers know the potential pitfalls that could delay or derail contract fulfillment. Exporting a single telecommunications satellite requires several Technical Assistance Agreements (TAAs), licenses issued by the State Department for the satellite's component parts. Congressional notification is required before a license can be issued, and State will not send up a license for review when Congress is out of session. Also, it can be difficult to segregate sensitive dual-use components. Civil, commercial, and national security satellites perform many of the same functions (such as communications relay, weather monitoring, high-resolution remote sensing) and share the same basic requirements for onboard systems (for example, transmitting and receiving antennas, data handling, power distribution, and thermal control). Inevitably, they will share common components which may be deeply buried in the system and thus are likely to be overlooked in a company's export control procedures. The difficulty extends to commonality with non-space hardware as well. In April 2006, for example, Boeing was hit with a $15 million fine for exporting guidance system chips that are used in missile programs. The tiny chips, costing less than $2000 apiece, were embedded in the guidance systems of commercial airliners costing $60 million each.

Foreign allies and friends openly expressed their resentment at being treated the same as potential adversaries in the wake of the 1999 export control reforms. That resentment, and the prospect of delays, denials, and added procedures prompted foreign manufacturers to seek alternative non-U.S. sources for space technologies. In some cases, U.S. components have been eliminated altogether. In July 1999, DaimlerChrysler Aerospace AG of Germany announced that U.S. components would not be used in its space hardware if similar components can be purchased elsewhere. Meanwhile, Thales Alenia of France and Italy has developed a so-called "ITAR-free" communications satellite, the first of which was launched on a Chinese Long March rocket in April 2005. This highlights one of the major unintended consequences of the export control policy change: instead of limiting international access to key space technologies, the policy prompted the establishment of a larger and more capable non-U.S. supplier base that can sell its products free of U.S. restrictions.

As U.S. satellite component makers shrink and lose market share, European manufacturers pick up the slack. CSIS reported that U.S. suppliers held a 90 percent share of the component market in 1995, but this dropped to 56 percent by 2000. Meanwhile, Europe's suppliers went from less than 10 percent in 1995 to 34 percent in 2000. In 1999 and 2000, U.S. firms imported one-third of the components used to build their satellites.

All of this adds up to a net loss in national security, contrary to the original intent. The combined effects of lost market share, a declining space industrial base, and deteriorating relations with allies and friends contribute to a long-term decline in the domestic availability of militarily critical technologies. Meanwhile, restrictions on U.S. firms on the selling of satellites and their components have not kept foreign countries, including China, from improving their indigenous space capabilities or from acquiring satellites or satellite services.

The export controls give foreign competitors a strong incentive to develop their own home-grown technologies, or at least to fulfill their needs with non-U.S. products. Ultimately, this will give the U.S. what space policy analyst Joan Johnson-Freese has called "100% control of nothing." In other words, U.S. influence over the spread of sensitive space technologies eventually will diminish to insignificance at the same time that the U.S. finds itself dependent on other nations for key components to its space systems. In the words of the 2002 CSIS study:

> . . . we may have done more harm than good for national security by restricting exports of replaceable technologies . . . continuing to treat communications satellites as munitions while we wait for some distant reform will lead to further shrinkage in America's satellite industry and expand the satellite industries of other nations.

CSIS produced another report on this subject in February 2008. A working group consisting of high-level representatives from industry and academia interviewed experts in U.S. industry and government as well as similarly placed authorities in Europe and Asia. They also had access to the results of a U.S. space industry survey conducted by the Commerce Department. The report reaffirmed previous findings that stricter export control had harmed U.S. market share and competitiveness, but "has not prevented the rise of foreign space capabilities and in some cases has encouraged it." Further, the report highlighted the disproportionate effect on the lower-tier contractors—those who supply big manufacturers like Boeing, Lockheed Martin, and Northrop Grumman—due to the high cost of compliance and the diminishing opportunities to team with foreign partners to bid on international projects.

The reports highlighted here are just a small sample of the many studies on this topic, all of which come to similar conclusions. This is not lost on the young people just starting their careers, who will be discouraged from seeking space-related jobs just as the baby boomers are retiring and the space industrial base is starving for talent.

By this time you're probably asking, "If everyone recognizes that the policy has failed—causing numerous unintended consequences while serving none of its intended purposes—why not just amend the law and fix it?" This is a reasonable question, but although more than a decade has passed, there is still not a satisfying answer. Policy-makers, including many who originally supported the stricter export control regime, realize that it has been counterproductive, but are reluctant to initiate changes for fear of appearing soft on national security. Of course, some policy-makers, as well as bureaucrats in the State Department and Defense Department, insist that there should be no backpedaling. To do so, some believe, would be tantamount to offering nukes to terrorist organizations or helping the Chinese build weapons systems to aim at the United States. One official told me there's no problem with the export regime, it's just that some U.S. companies are too stupid or too lazy to properly fill out their export license applications. Aside from grossly oversimplifying the scope of the problem, this point of view also neglects the fact that some U.S. subcontractors are too small to have a team of lawyers devoted to export control paperwork.

This problem affects more than just satellite builders and component makers trying to cut business deals. A key element enabling space exploration efforts will be NASA's ability to share technologies and information with its contractors and international partners, and to locate and absorb useful technologies from a wide variety of sources. However, strict export control laws stand in the way of establishing smooth two-way flows. Reports in trade publications and interviews with European officials have noted that despite satisfaction with trans-Atlantic projects such as the Hubble Space Telescope and the Cassini-Huygens mission to Saturn, Europeans lament that cooperative ventures such as these are no longer possible under the current U.S. export control regime. If this perception persists, it will present a major obstacle to the building of international alliances for exploration projects. NASA will need to obtain waivers of export restrictions for exploration partners on space hardware, software, and information.

All of the consequences of the tighter export control policy—ineffectiveness at restricting the spread of technologies, loss of U.S. market share, increase in foreign investment in competing products, further damage to an already ailing U.S. space industrial base—were forecast by experts prior to initiation of the policy shift. This important component of the nation's technology policy fell victim to shortsighted decision-making practices identified in Chapter 3: haste, poor planning, and a desire by the Congress to assert authority and make a political statement.

An important lesson to be learned from this episode is that once a bad policy is put in place, it can be very difficult to fix it, even if there's a consensus that it needs fixing. Since the Obama administration took office, the political environment for turning around the export control problem is better than it's ever been, but it will still be a long, slow process. So be careful what you wish for—you might get a short-term gain, but the cost will be long-term pain.

Chapter 7
Earth as an Open System

We cannot solve our problems with the same thinking we used when we created them.—Albert Einstein

The discussion in Chapter 1 noted that the extensive literature on global problems and the search for solutions tends to assume the Earth is a closed system—nothing goes in or out except light. This is a very odd view, considering that NASA and its counterparts around the world have spent the past half-century demonstrating that it's not true. Gases escape into space from the fringes of the atmosphere. Satellites are launched into long-lived orbits and deep space probes depart the Earth, never to return. Granted, these constitute a miniscule fraction of our planet's mass. For something a little more substantial, let's consider what's coming in: an estimated 40 million kilograms of meteoroids enter our atmosphere every year. Fortunately for humans so far, they tend to arrive in small pieces that disintegrate before reaching the ground. Also fortunately for humans, they may have brought the seeds of life to this planet about three to four billion years ago.

If people are more comfortable dealing with a closed system, then we should be talking about the solar system, not the Earth. Of course, that's not accurate either—the Pioneer and Voyager space probes are exiting the solar system, as is the solar wind, and who knows what's coming in—but it will suffice for our purposes. The most prominent feature of our neighborhood is the fusion furnace at the center, which has a diameter 110 times that of Earth, mass 330,000 times greater than Earth, and luminosity reported to be 390×10^{18} megawatts. That's an unimaginably large power source, and only an infinitesimally small fraction of it has been tapped by humans.

Astronomers have joked that if alien surveyors were to pass through our part of the galaxy, they would catalog us as a single-star system with four planets (Jupiter, Saturn, Uranus, and Neptune) and assorted debris (including Earth). This suggests that anyone who needs a lesson in humility should study astronomy. For those who have some notion of our place in the cosmos but are still bold enough to venture beyond the little mote of dust we call home, there's plenty of other debris nearby that we may find interesting and useful. To give just one example, the total mass of the main belt asteroids, though considerably less than the mass of Earth, is something close to 300 x 10^{19} kilograms, a respectable amount.

The pioneering Russian space theoretician Konstantin Tsiolkovsky remarked that Earth is the cradle of mankind, but one does not live in the cradle forever. This is a reasonable statement if you consider that travel times to solar system destinations, even using the primitive propulsion methods available to us today, are measured in months and years, mere fractions of a human lifespan. Is that really much of a stretch for humans, given the comparable amounts of time we spent on dangerous ocean voyages and investigation of uncharted continents during the European age of exploration? We may go beyond the solar system someday, but I'll leave that to the science fiction writers. Restricting ourselves to our own solar neighborhood may seem like thinking small, but let's take this one century at a time.

Before we pack our bags and board our spaceship, we need to answer the "why" question. We're talking about an investment of time, talent, and wealth far beyond anything previously contemplated by humans. We'd better have a rock-solid rationale, and the courage and determination to stick with it. The explorers of old ventured out to seek new wealth, claim new lands, and improve their condition in other ways, such as achieving religious freedom and escaping tyranny. Things like scientific discovery and fulfilling the human drive to explore were undoubtedly motivators for some individuals, but these were secondary objectives, not the goals that justified the enterprise on a societal level. Have we established a clear, compelling rationale for our movement into space?

When voices of authority and advocacy, such as NASA, the Congress, and space interest groups, talk about a long-term strategy for human spaceflight that's worth the investment and the risk, we hear generalities about national pride, human destiny, international cooperation, and inspiration of youth. There's also the argument that it will give the U.S. space industrial base something to do, which sounds more like make-work than a noble cause. To the explorers of previous centuries, these would all sound like secondary benefits, and mostly short-lived ones at that. Even collectively, they don't add up to a primary rationale for engaging in something of this magnitude and risk under the circumstances likely to be faced by humanity in this century. Maybe they're sufficient justifications for individual advocates and those who would directly benefit from the work, but not for the whole of society.

If we're serious about moving out into the solar system, exploration and development need to go hand-in-hand. Exploration won't get very far before being forced to turn back unless development follows in its wake. Exploration brings excitement, intellectual challenge, and for some even spiritual fulfillment, but development brings two absolutely essential elements: relevance to society, and the means to continue.

If all the resources of all the space agencies of the world were combined, they would still be inadequate to the task of exploring and developing the solar system. Other sectors of society need to participate, and new wealth and capabilities must be created along the way. For this to happen, we need to change the short-term incentives in our institutions and our way of thinking about the long-term future. This will take time, but the work of refining the rationale for spaceflight needs to be done without delay.

Wrestling with rationales

On May 3, 2006, NASA's then-administrator Mike Griffin gave a talk to a breakfast meeting on Capitol Hill sponsored by the professional association Women in Aerospace. I had a meeting on the Hill that morning along with a colleague who was attending the breakfast. I met her as she came out and asked her what she thought of the Administrator's talk. She shook her head in a manner that indicated dismay, or puzzlement, or perhaps a combination of both. Griffin had given an enthusiastic description of the space transportation architecture that NASA was planning for the return to the Moon, and followed it by saying, "We haven't got a clue what we're going to do once we get there." Griffin's remarks weren't posted on NASA's website with his other public speeches, so I don't know whether this comment was planned or off-the-cuff. In either case it does not bode well for NASA's ability to convey its purpose to the public—to answer the "why" question—if the agency's top official is unable to do so.

In contrast, just a few weeks before Griffin's breakfast talk, the American Astronautical Society's annual Goddard Memorial Symposium featured a speech by John Marburger, the president's science advisor. In an oft-quoted remark from that speech, he addressed the question of the long-term rationale for spaceflight by saying that it boils down to "whether we want to incorporate the Solar System in our economic sphere, or not... at least for now, the question has been decided in the affirmative." This may have been a bit of a stretch since Bush administration policy simply recognized that the U.S. has economic interests tied to space. Nevertheless, it was a forward-looking comment that got to the heart of the spaceflight rationale more than any other statement from that administration. It encapsulates an important aspect of where we should be aiming.

The rationales for space exploration that we've been hearing—national prestige, scientific discovery, technological spinoffs, inspiration of youth, and our "destiny" or "nature" to explore—are the same arguments that were used to justify the value of the Apollo program. They were more than sufficient at that time, but this is a very different time. In the half-century since the launch of Sputnik and the creation of NASA, a great deal has changed, from international economic and geopolitical relationships between countries, to the proliferation of space technology throughout the world, to the circumstances of growing up in hometowns (and wired households) in America and other developed and emerging nations.

All the rationales mentioned above remain good, but are no longer sufficient. Although space offers an array of worthwhile secondary rationales, nationally and globally we still need better agreement on the primary rationales. There is also the question of whether we can unambiguously achieve all of these objectives as effectively as we did in the 1960s with the Apollo program. In the economically globalized and more technically advanced environment of the 21st century, making such a case is far less credible today. So before we automatically embark on Apollo 2.0, we should reconsider each of them.

National prestige. The international reputation of the United States has changed considerably since the Cold War years, and in many ways it has suffered. Perhaps it was inevitable that the "great power" emerging victorious after decades of superpower confrontation would become the target of so much resentment for all the world's problems, whether or not it's deserved. This can be exacerbated by other factors. If the U.S., through its statements or actions, puts forth an image of "sole superpower," and seeks to promote that image by nationalistic space achievements, proponents of globalization will see this as clashing with the political, economic, and technological leveling effects of globalization that compel the rest of the world to view the concept of "superpower" as undesirable and archaic. On the other hand, the mostly inaccurate view of many anti-globalizers is that globalization is the Americanization of the world, and that this must be stopped because it damages other cultures. Such perceptions manifest themselves in many ways, including trade disputes, disagreements over environmental issues, rejection of diplomatic initiatives, and terrorism. The result is that both supporters and opponents of globalization have reason to resist attempts by the U.S. to flex its nationalistic muscles.

It seems highly unlikely that U.S.-dominated exploration of the Moon and Mars, no matter how successful, would win many hearts and minds in the international arena unless large-scale benefits to Earth (not just the U.S.) were clearly visible as a result. Otherwise, the likely response around the world would be, "They should have used the resources to [cure AIDS, end starvation in poor countries, address global climate change, *fill in your favorite world problem*]."

Science. Much has been said and written about the wealth of knowledge that the space program has brought to us about Earth and the rest of the universe. The past half-century has seen textbooks rewritten multiple times and the emergence and blossoming of new scientific disciplines. Debates about programs and funding have always sought the right balance between human and robotic missions that will produce the most value for our science dollars. The question that needs to be addressed in current planning for human exploration is whether science is a primary or secondary goal. If it's a primary goal, there will be plenty of work for people to do on the Moon and near-Earth asteroids through mid-century, but not elsewhere because sending robots will be a far more efficient use of resources. In other words, as the sophistication and productivity of robots improve, there is no scientific motive for a rush to send humans to Mars and beyond that justifies the added risk and expense. Having scientists working on site is the preferred approach when the cost-benefit analysis makes sense. It's reasonable to believe that this will be the case on the Moon in the foreseeable future, but we lack the knowledge and experience to make a credible estimate of when this will be true of Mars and other more distant destinations.

Alternatively, science may be a secondary goal in humanity's expansion into space, in which case robotic investigation is still the preferred approach for science missions. Making science a secondary goal, as it was in the Apollo program, does not imply that it isn't important. Like many in the Apollo generation, I have loved science ever since I learned to read. In fact, in my early years of literacy I thought that the main purpose of reading was to learn about science. For me, science is a powerful justification for sending robots and humans into space. But I'm just one American taxpayer among a couple of hundred million, and I'm certainly not your average guy on the street when it comes to my level of interest in this topic. Joe and Jane Citizen, who are paying the freight for this activity, want to know, "What's in it for me . . . for my country . . . for Earth? And what's it going to cost me?"

Technology spin-offs. It's true that high-tech spin-offs have brought benefits to society and boosted the U.S. economy, although the value of such benefits is impossible to measure with any precision. However, spin-offs are not a sufficient justification for a space exploration program. They are secondary benefits, and an investment of this magnitude must be justified on its primary benefits. Any attempt to argue that spin-offs provide the rationale for the program is easily countered: direct investment in technology development in the absence of a space program would bring similar results at less cost.

The old saying, "necessity is the mother of invention," has been quoted often to make the argument that numerous technological innovations would have appeared much later, if at all, were it not for the investment made in the space program. There's some truth to that, but it's a much harder sell than it used to be. By the 1980s, U.S. industry had surpassed the government in technology investment. It's hard to imagine that technical advances would move any more swiftly in many key areas just because a Moon-Mars exploration program is underway, especially considering the modest funding levels being considered and the preference for use of proven rather than cutting-edge technology. Computer hardware and software, telecommunications networks, medical technologies, and most other engineering advances are now being driven by private sector and non-space government investment. However, one area where the needs of space exploration could significantly increase the pace of development is robotics, since physically demanding operations in remote, hostile environments are especially suited to robotic solutions.

Inspiration. Administrator Griffin's speech noted earlier was not the only occasion in which NASA's top management failed to compellingly articulate the rationale for sending humans back to the Moon and on to Mars. NASA's Associate Administrator for Exploration Systems at that time, Scott Horowitz, gave a speech at a Capitol Hill luncheon sponsored by the Coalition for Space Exploration on September 14, 2006. Horowitz, a former shuttle astronaut who is called "Doc" by his colleagues, exudes a real passion for his work. His answer to the "why" question: It will inspire our youth. Current and future generations, he believes, will develop a burning desire to explore space and will flock to engineering, math, and the sciences, just as they did during the Apollo era.

One hopes that Doc Horowitz can think of more reasons to explore space. Inspiration to youth is a very positive *side-effect* of the space program, but it's not a primary rationale for going into space or a justification for expending massive resources and taking on exceptionally high risk. Another problem with Horowitz's statement is that post-baby boom generations will not respond in the same way and to the same degree as he and I did when we grew up watching Project Apollo unfold.

Over the years, I've heard many in the space community, including Horowitz, invoke the dinosaurs-and-space myth, which goes something like this: All kids are fascinated by dinosaurs and space. Dinosaurs are part of the prehistoric past, while space is part of the present and future. Therefore, all kids are space enthusiasts who see space as an important part of their future.

There is no data indicating that the dinosaurs-and-space myth has ever been anything more than wishful thinking on the part of the space community. Even if it was true in the past, that era ended by the time of the 1986 *Challenger* accident, making it a bad assumption for the planning of NASA's education and outreach programs. In contrast to kids growing up in the Apollo era, young people today have many more information sources and distractions: 100-plus television channels, computers linked to the Internet at home and at school, cell phones and portable entertainment devices, and far more extracurricular activities at school than were common in previous generations. A downside of this great abundance is information overload—a fact of life for people of all ages, but more challenging for young people who have not yet developed the judgment to filter it properly. Space achievements have a difficult time emerging above the noise level. Even when they do, the attention they gain may be fleeting as young people move on to the next thing rather than seek out more knowledge about space and its real and potential effects on their lives.

Another consideration is that enthusiasm for space does not necessarily translate to support for real-world programs. Space entertainment—consisting of movies, TV shows, websites, video games, science fiction books and periodicals—has enjoyed considerable popularity in recent decades, but its avid followers may have little or no interest in real spaceflight and space science programs. To assume that this group is already on board as active supporters would be a mistake. In any case, little is known about whether they tend to be politically active, or would be inclined to become so.

Destiny. In a series of essays posted on NASA's website in 2004, the agency's chief historian, Steven Dick, discussed "Why We Explore." He asserted that "we cannot afford NOT to explore" because "no nation can long afford to sacrifice long-term goals for short-term needs." I can certainly agree with that. As you may have noticed by now, the importance of taking the long-term view is the central theme of this book.

History gives us many clear examples of how exploration has played a key role for certain individuals and cultures. On the other hand, it's equally clear that not all individuals and cultures embrace exploration. Tribes in some parts of the world have stayed within the confines of the same region and have lived a mostly unchanging lifestyle for hundreds of years. Even large, modern societies have many members who have no desire to stray more than a few hundred kilometers from where they were born. One of Steven Dick's online essays quoted historian Stephen J. Pyne, who argued that:

> Exploration is a specific invention of specific civilizations conducted at specific historical times. It is not . . . a universal property of all human societies. Not all cultures have explored or even traveled widely. Some have been content to exist in xenophobic isolation.

In other words, we're not born explorers, at least not all of us. This weakens the argument that as individuals or localized cultural groups we're destined to reach for the stars. But what about larger groups like nations or the entire human species?

If a society is to grow, enrich itself, advance its technology, and simulate its creativity, it must explore in some manner. That doesn't necessarily mean space exploration will be the first choice, even if the technological capability to do so exists. A multitude of mysteries await exploration without looking beyond Earth, from the microscopic world to the deep oceans to the human brain.

Reportedly, when early 20th century British mountaineer George Mallory was asked why he wanted to scale Mount Everest, he responded, "Because it is there." People eagerly support adventurers like Mallory, cheering them on and rejoicing if they succeed. (Mallory didn't. He died on Mount Everest in 1924, and his body wasn't found until 75 years later.) However, when the adventure is several orders of magnitude more risky and expensive and is being paid for by their business investments or tax dollars, people feel compelled to seriously consider the trade-offs and demand more substantial justification. Analysis of opportunity costs is inevitable: if we invest substantial resources in space, what other investments are we sacrificing? Are we making sensible trade-offs? Is space development a societal undertaking, an entrepreneurial venture, or a mix of both?

Another way to look at the rationale for space is to compare it to other great human endeavors throughout history. Neil DeGrasse Tyson, an astrophysicist and director of the Hayden Planetarium in New York, has taken this approach. Tyson believes that great things—which he defines as "expensive or audacious" endeavors like the Apollo program, the Manhattan project, the pyramids, the Great Wall of China, and the cathedrals of Europe—are the result of three drivers: national security, economics, and praise of deities or royalty. He concludes that human expeditions to Mars and beyond certainly qualify as expensive and audacious, so if they don't satisfy at least one of the three drivers, we're not going. It's evident that two of these three are non-starters, so we need to rethink and expand our conceptual approach or else we're not going. Let's examine each of Tyson's drivers.

Looking ahead at least through mid-century, there is no evidence that a direct national security benefit will result from exploration and settlement of the solar system. The United States already has a very substantial national security space program that has no requirement for people in deep space. Someday, there may be a global effort to plan for *planetary* security, in which case the aim will be to prepare for interception of a threatening object headed for Earth. Diversion or destruction of such an object is preferably done as far from Earth as possible, which may mean it will be done by automated spacecraft. Planetary security will be one of our primary rationales as the capability to do something about it becomes available—or if an impending threat appears—but it doesn't provide an immediate driver. Despite the potentially catastrophic consequences for humanity, the impact of a large asteroid is a low-probability event. Our mindfulness of planetary security is not even close to the way we think of national security, which is in the news every day and entails a combination of military capabilities, intelligence functions, national prestige, and diplomacy affecting geopolitical relationships here on Earth.

As for Tyson's other drivers, we're clearly not venturing out into the solar system to praise deities or royalty, and even if we wanted to, I can't imagine we'd ever be able to agree on which ones to praise. So that leaves economics—and an expanded interpretation of planetary security.

Economics as a primary rationale

I've seen pictures of the martian surface taken by NASA rovers that have been altered to include business franchises such as McDonald's and Wal-Mart. Cute, but not likely. Whenever I see these pictures, I note that the parking lot is empty, which would seem to indicate a flaw in the business plan. Even the addition of a multiplex theater and a Starbucks would fail to generate sufficient retail traffic.

The economic focus of space development initially needs to be on direct benefits to the Earth-based economy—marketable materials, energy, and unique products, not just technology spin-offs—and eventually on expansion of space infrastructure to extend our reach and our capabilities.

Chapter 5 pointed out the immeasurable economic benefits that have resulted from space technology applied in three areas: communications, navigation, and Earth observation. These space applications will continue to develop in the 21st century, improving their precision, adding services, and extending their reach more broadly throughout global society. But it would be surprising if these continued to be the only space applications contributing to national and global economies. We should see the emergence of useful new capabilities in areas that until now have been relegated to hype or science fiction. Examples include:

- Development of microgravity materials processing techniques to produce better products such as metal alloys, crystals, optics, and pharmaceuticals.
- Collection and distribution of energy from space for the terrestrial power grid and other applications. Like communications signals, beamed energy is a weightless electromagnetic product that has near-universal demand on Earth, and no carbon footprint. Relay satellites also may be used to transmit power from point-to-point on Earth, allowing redirection of surpluses to areas in need.
- Routine construction and repair in space. As cislunar space (the area within the Moon's orbit) becomes more active, infrastructure elements will be built and repaired there, with only the

highest-value components shipped up from Earth. A variety of plug-and-play platforms will provide affordable space services, and a used satellite market will develop.
- Extraction and processing of extraterrestrial materials, from the Moon and other near-Earth objects, to support continued development of space infrastructure. As space facilities and experience grow, transport of processed materials down to Earth will become economical. Lightweight, high-value products will dominate the early years of this activity, but eventually the increasing efficiency of space-to-Earth transfers may allow bulk products to be transported economically as well.
- Eventual movement of industrial activities off the Earth to alleviate damage to the terrestrial environment and to take advantage of proximity to the raw materials, energy, and hard vacuum available in space.

A long-term vision should target tangible benefits, as explorers always have done, and as investors, governments, and constituents always demand. It would be a strategic error to equate destinations with long-term goals, as too many space advocates have done. The next phase of human spaceflight should be a quest for valuable capabilities that in turn drive the choice of destinations, not the other way around.

At our early stage of development, it's difficult to determine the appropriate scope and rate of human movement into space. Scott Pace, director of the Space Policy Institute at George Washington University, has developed a way of framing this analysis by asking two fundamental questions about the future of human spaceflight: Can humans "live off the land" in off-world settlements? Are there economically useful activities in space that can sustain human communities? He constructed a simple matrix using terrestrial analogies to help address these questions.

	Nothing commercially useful	*Commercially sustainable*
Live off the land	Antarctica	Human settlements
Can't live off the land	Mount Everest	Offshore oil platforms

Although Pace's approach isn't designed to encompass all the details necessary to make critical strategic decisions, it does cultivate a way of thinking about the situation that is important for capabilities-driven planning. If humans can live off the land and do something commercially sustainable, then settlements can thrive. If they can't do either, then humans will make symbolic visits and do a little science. If they can live off the land but find little or no commercial benefits, long-term scientific outposts are a plausible result. If there's commercial payoff, but humans can't live there for long, operations will be human-tended but predominantly robotic. This is a different way of thinking than was prevalent in the early days of the space age.

Project Apollo was the right thing to do in the 1960s but it taught us the wrong lesson about the future of space exploration. The history-making success of Apollo's destination-driven model appeared to demonstrate the right way to approach the human movement into space, and this experience continues to affect the thinking of many people today. In fact, Apollo is the wrong model—an unusual effort that occurred in unusual circumstances. President Kennedy, who considered other options before picking the Moon landing, did not choose the Moon for its value as a destination. His goal was to demonstrate an advanced technological capability that would dramatically surpass the Soviet Union in the eyes of the world.

Politically and technologically, Apollo was an undeniable triumph. But the goal was very limited in both scope and duration: put a man on the Moon and return him safely to Earth before the end of the decade. Neither the stated goal nor the intent behind it provided for the establishment of a permanent space infrastructure or any other plan for the movement of humankind into space.

When Apollo ended, it left a disappointing legacy. The remaining infrastructure included some launch facilities, tracking stations, and NASA research centers, all underutilized in the aftermath of the program. Leftover flight hardware was put in museum exhibits, and no new systems with the ability to go to the Moon were being seriously considered. The U.S. space industrial base was gutted, as a large percentage of the scientists, engineers, and technicians who designed and built the program were laid off, many of them never to return to space-related work. NASA's shrinking budget prompted a reduction of almost 40 percent in civil service personnel and nearly 400,000 contractors.

In a way, the race to the Moon was like a person on a diet attempting to lose 20 pounds. For a time, a dieter accepts the rigors of the challenge and eventually produces the desired result. But the diet is a deviation from normal behavior and a change in resource consumption. When the goal of losing 20 pounds is reached, the dieter goes back to normal behavior and the weight comes back. That's why diets don't work in the long run—they are anomalous behavior that can't be maintained. To achieve a lasting change in one's health and fitness level, one must create a "new normal." The same is true for a healthy space program. An active space enterprise that brings ongoing benefits to Earth and continuously builds on its capabilities must become the "new normal" for U.S. and global space efforts.

Some may argue that a capabilities-driven approach, especially one seeking payoffs on Earth, is too inward-looking, or uninspiring, or simply lacking in vision. I believe that in the context of 21st century space efforts, the opposite is true: continuation of the destination-driven approach, which has dominated thinking for a half-century, is a persistent non-vision we cannot afford. It fails to take a decades-long perspective that adds lasting value and progressively builds sustainable infrastructure and communities. As a result, it hinders rather than helps the movement outward. Programmatically, human landings on the Moon and Mars are treated like the finish line in a race, and planners have insufficient motivation and resources to think beyond that point. If society's stakeholders ask what happens next after we reach the finish line, as they inevitably will, the answer needs to be something better than "we go back again" or "we move on to the next destination."

To highlight the lasting importance of efforts to develop space infrastructure for economic and social benefit, it's instructive to look back at President Kennedy's famous address to a joint session of Congress delivered on May 25, 1961. Not the part that everybody remembers in which he called for a manned mission to the Moon; rather, let's focus on two short statements that are less well-known but arguably have proven more far-reaching than the celebrated words about the Moon race. Kennedy asked the Congress for additional appropriations to "make the most of our present leadership, by accelerating the use of space satellites for world-wide communications," and to "give us at the earliest possible time a satellite system for world-wide weather observation." The result was a much quicker pace of deployment for these capabilities, and just as importantly, their availability to a much wider audience than envisioned previously.

Three decades later, after communications and weather satellites had proven their worth to the point of becoming indispensable, the

U.S. government presented another gift to the world in the form of satellite navigation, giving free access to the Global Positioning System constellation. The resulting leaps in human welfare and potential just from these three space applications have been truly exciting, even though the space systems themselves tend to be invisible to the users.

Today's space services facilitate growth and enable the detection and analysis of regional and global-scale problems. Communications and navigation have increased the speed, efficiency, and safety of doing the business of corporations, governments, and individuals. Earth observation, as noted in Chapter 1, helped keep the Cold War from becoming a hot war, and since the late 1970s, has discovered, confirmed, or otherwise measured much of what we know about climate change, which will help us address a different kind of threat to our existence. The next stage in space development must continue the tradition of contributing to global solutions because, in the words of historian Paul Kennedy, "it is not enough merely to understand what we are doing to our planet, as if we were observing the changes through a giant telescope on Mars."

Survival as a primary rationale

The survival rationale for space development is usually associated with two related concepts: planetary protection against space object impacts, and preservation of the human species through interplanetary migration. In the first case, the idea is to devise ways to divert or destroy incoming asteroids; in the second case, the assumption is that something catastrophic may happen to Earth, so it would be desirable to have viable human communities, and storehouses of human knowledge, in other locations. To this we should add a third case with a higher probability of near-term consequences: planetary stewardship, in which Earth faces potentially calamitous terrestrial threats and needs space capabilities to aid in mitigation and adaptation.

A civilization-ending asteroid impact is, thankfully, a very low probability event. As such, it fails to stir the fear and sense of urgency that the threat of nuclear attack did during the Cold War. Similarly, the need to become a multi-planet species is viewed by most people as a very low priority, or at least as something that lies several generations into the future. In both cases, there's almost no chance of sustaining the focus of decision-makers, institutions, and constituents, and maintaining the required flow of society's resources in the face of competition from more immediate needs. Based on these two cases, survival doesn't look very compelling as a primary rationale if the intent is to generate significant forward motion in this century.

On the other hand, terrestrial threats that are global in scope are clearly in evidence today and will have serious consequences in the near—to mid-term future. Climate change, environmental degradation, and resource depletion are obvious problems for which current and future space applications should be part of the solution set. Throughout most of human history, when problems like these emerged—the weather got too cold, the river dried up, the farmland stopped producing crops—the solution was to move somewhere else. Today, that doesn't work because there are too many vulnerable areas with too many people whose migration, within and across national borders, can cause severe stress and even violent conflict. Climate and environmental disasters have the potential to affect at least as many people as the nuclear conflict we feared during the Cold War, a time when our security strategy consistently planned for a wide range of contingencies, including worst-case scenarios. It would be wise to do the same in the 21st century, and as before, to make the best possible use of space capabilities. (The next chapter provides some suggestions.)

The survival rationale in the 21st century timeframe does not include migration to permanent space settlements as a feasible solution to the world's population pressures. The U.S. Census Bureau estimated that as of 2009, global population was experiencing a net annual increase of over 79 million, which is more than nine times the size of New York City. Even at the lower growth rates expected in mid-century the annual net increase will still be around 50 million. I don't believe we'll be able to transport tens of millions of people per year to space cities at any time in this century. If we learn how to construct and operate large off-world ecosystems and migrate hundreds or perhaps thousands of people there by the end of the century, that will be a truly impressive achievement. Meanwhile, population problems on Earth can't wait for a space migration plan. We need to find ways for space applications to contribute to the resource and lifestyle needs of the multitudes who will be remaining on Earth for the foreseeable future.

For the remainder of this century, planetary stewardship could look something like this: Earth would no longer be treated as a closed system. Advanced space applications would relieve some of the stresses that humanity has imposed on our planet. Its ecosystem would recover and be preserved through increased use of extraterrestrial resources that relieve the need for excessive ground-based generation of energy and the extraction, transport, and processing of raw materials. As land use patterns change, and space technology provides ubiquitous services (including high-quality education and telecommuting opportunities) that are not reliant on major population centers, the current incentive to abandon the countryside and flock to big cities would be reversed. Just as the automobile determined settlement patterns in the United States during the 20th century, space technology would be an integral driver of settlement patterns worldwide in the 21st century.

Framing the rationales

If the most popular rationales for space development—national prestige, scientific discovery, technology spin-offs, inspiration, and human destiny—are actually secondary goals, what argument can be made that economics and survival deserve to be the primary justifications for spaceflight in the 21st century? Helpful guidance comes from the work of psychologist Abraham Maslow. Beginning with a 1943 journal article titled *A Theory of Human Motivation* and developed further in articles and books in the years that followed, Maslow crafted his Hierarchy of Needs, which is graphically depicted as a triangle divided into five levels, with the most mundane at the bottom and the most cerebral at the top. According to Maslow's theory, a person must satisfy their needs at a particular level in order to move up to the next one.

The base level consists of *physiological needs*, which are those required to sustain life such as air, food, water, and sleep. The next level is *safety needs*, which involve freedom from the threat of physical and psychological harm and may include things like shelter, basic security, and health care. Having achieved that, we move up to *social needs*, those related to interaction with other people such as group belonging, friendship, and love. Those relationships allow us to go a step higher to *esteem needs*, in which we seek self-esteem and a sense of accomplishment (internal motivators), as well as recognition and social status (external motivators). Finally, at the summit of our existence is

Maslow's Hierarchy of Needs

self-actualization, where we reach for our full potential through efforts that are never completely satisfied as we search for truth, justice, wisdom, beauty, or in the words of author Douglas Adams, the meaning of life, the universe, and everything.

The difficulty with the popular rationales for space is that *they try to appeal to self-actualization, the highest need level*. Concepts like national prestige, scientific discovery, inspiration, and destiny are worthy things to aspire to, so most people will acknowledge their support. But these things exist at a need level that most people haven't reached. That's why public opinion polls show broad support for space, except when people are asked to prioritize it against social programs, national defense, law enforcement, environmental protection, or other more basic needs.

It could be argued that technology spin-offs appeal to different need levels depending on their nature. High-tech gadgets are consumer items that can affect self-satisfaction and social status at the (still very high) esteem-needs level, while medical spin-offs could reach down to the safety-needs level. But this should not cause us to forget that spin-offs, by definition, are secondary applications, so they can't be a primary rationale, and in any case most people will fail to associate them with space investments.

Fans of the Apollo program may be quick to respond: "Wait a minute—if what you say is true, then why were those rationales sufficient to justify the race to the Moon?" The answer is that different circumstances call for different approaches to need fulfillment. As NASA Administrator James Webb and Secretary of Defense Robert McNamara suggested in 1961, the space program was "part of the battle along the fluid front of the Cold War." In terms of Maslow's Hierarchy, national prestige was not a self-actualization concept at that time—it found a place on the second level, safety needs, as much of American society associated the space program with the arsenal of weapons fighting communism. To a great extent, the same was true for science and inspiration. Even the young, who were less aware of the geopolitical threats at that time, saw science and inspiration at least as far down the Hierarchy as the social-needs level, where the real-life adventure of spaceflight promoted friendships and group bonding through common interests in astronauts and rockets.

Active members of the space community today will have a different perception than the rest of society regarding where the rationales fall in the Hierarchy. To the people who live this stuff on a daily basis, the exploration and development of space is their financial security and medical coverage at the safety-needs level, their professional and social network at the social-needs level, and their recognition and sense of accomplishment at the esteem-needs level. Many have reached the self-actualization level where they seek the truth about life elsewhere in the universe, marvel at the beauty of an image from the Hubble Space Telescope, and delight in elegant solutions to engineering challenges. The space avocation is satisfying on so many levels, it's no wonder that those who are involved in it want to keep doing it, and are surprised when they encounter people who don't recognize its power to bring fulfillment.

The rest of the populace outside of the space community needs to see how space contributes to need fulfillment at the levels of the Hierarchy where they live and work. That's why economics and survival, as depicted above, can serve as primary rationales. They may not be good candidates for self-actualization, but they show up at every other level of the Hierarchy. Taking a long-term view that factors in climate change and other global threats, they can even reach the physiological-needs level if food, water, and other resources become more scarce and space can contribute to a solution. For those who want to go beyond the first four levels and seek self-actualization through space activity, the path to the highest level must traverse the lower levels by employing the economics and survival rationales.

It's clear that the space program needs to throttle back on its long-cultivated image of being something "special," which today comes across (in Maslow's terms) as serving self-actualization needs. The "space is special" approach was perceived differently and worked well during the early years of the space age, when spaceflight was new and the space community was a small, elite group. It still works for lifelong space geeks like me, for whom phrases like "galactic cluster" and "space industrialization" can trigger sweeping vistas of the mysteries and treasures awaiting us as we venture out from the home planet. That's because space was opened up to humanity for the first time as my generation was growing up and coming of age. We watched it happening, and we embraced it. I was born three years before NASA was created, and I lived just a couple of miles down the road from the agency's Lewis (now Glenn) Research Center in Cleveland, Ohio. I remember watching the addition of new buildings to the Center in the 1960s. Despite the frenetic pace of the construction, new employees were arriving faster than they could be accommodated, so they set up temporary office space in a storefront one block from my house, with the big NASA "meatball" logo proudly displayed. This made a great impression on a junior space cadet like me, because NASA wasn't just a collection of offices filled with government bureaucrats, it was a place where science fiction was being turned into reality. As Arthur C. Clarke once said, any sufficiently advanced technology is indistinguishable from magic. To me and many others in those days, NASA was doing magic.

That was then, this is now. For later generations, or even for those in my generation who never earned their space cadet badges, the "space is special" approach to politics and public outreach is no longer helpful and may be harmful. Occasional episodes of interesting and exciting space activity will have their moments in the news cycle, but must be seen in the context of other national and global interests. A "special" image may invite competition from other interests seeking to assert that space is no more special than they are. For space activity to prosper, it needs to behave, and be widely perceived, as a mainstream activity (albeit a very cool one) that benefits the national interest and contributes to basic personal needs—not as an expensive luxury or a mid-life crisis project for baby boomers.

Proponents like to think that space has broad, enduring support, and point to survey results like the Gallup polls conducted since 2005 that were commissioned by an advocacy group called the Coalition for Space Exploration. Like other polls over the previous three decades, these more recent Gallup polls revealed that an impressive majority of respondents "supported" or "strongly supported" space exploration. That's reassuring, but don't read too much into it. The responses tended to be dominated by baby boomers old enough to remember the Apollo era. Respondents were not asked to prioritize the space program against other government programs or societal needs. And in a January 2009 survey, they were enticed with exaggerated information about space-related economic benefits. (For example, respondents were presented with U.S. aerospace sector employment numbers and trade surplus figures and led to believe these were all attributable to space, which only makes up a fraction of that sector. Also, a variety of common items, from cordless tools to joysticks to vacuum cleaners—yes, *vacuum cleaners*—were listed as space spin-offs.) The designers of Gallup's January 2009 survey clearly understood the importance of the economic rationale, but unfortunately used it in a misleading way that doesn't give us a reliable reading of what people would think if they had good information and needed to prioritize the expenditure of society's limited resources.

Elsewhere we find evidence that spaceflight is still perceived by many in the public as an optional activity, not necessary for their economic benefit or survival. To cite one example, a Harris poll on fixing the U.S. budget deficit conducted in March 2007 asked respondents to pick two federal programs (from a list of 12) that should be cut to reduce government spending. The space program was chosen by 51 percent of respondents, topping the list by a 13-point margin above the second most popular choice. This snapshot of public opinion suggests that approximately half of the U.S. voting-age population views the civil space program as either a waste of resources or simply a non-essential activity.

Multiple-choice surveys can lead people in certain directions depending on which questions are asked, how they're worded, and in what order they're asked. If it were practical to conduct a survey using essay questions, it would be instructive to ask people to write a couple of paragraphs to complete this thought: "I believe space exploration and development is (or is not) an important part of our future because . . ." What would be the distribution of pro/con responses? How many people on either side would offer coherent, well-informed responses? How many would simply say "Don't know"? Just as importantly, how many of the good responses would focus primarily on economics and survival rationales, as opposed to the popular self-actualization rationales like prestige, science, inspiration, and destiny? Maybe we should require this essay as a homework assignment in high schools and universities across the country so we can learn what upcoming generations are really thinking.

Priorities and choices

In a lecture at George Washington University on April 29, 2009, former presidential science advisor John Marburger made the case that the nation needs more and better of science policy analysis because in the absence of good, thorough analysis, advocacy trumps science. That may seem like an unremarkable observation about a situation that has parallels in every sector of public policy. But Marburger emphasized this point because for decades, despite the critical importance of science and technology (S&T) to modern society, policy analysis in support of important decision-making has been too often poorly executed, or ignored, or nonexistent. Some S&T topics, like medical research and the environment, have gotten more attention than others, but space has garnered relatively little interest. A friend of mine who spent many years at NASA Headquarters likes to joke that the nation's space policy community is 79 people who all know each other. That's closer to the truth than we'd like to admit.

Space policy analysis is needed to provide an array of options to decision-makers. Beyond simply listing feasible options, skilled analysts must bring to bear their knowledge of history, politics, international affairs, technology, and futurism to dive deeply into the positive and negative implications of each option and to project outcomes into the future. As Marburger warned, in the absence of thorough analysis, decision-making will be driven by advocates who may be program officers, lobbyists, or others who are pushing a particular solution. This tends to stifle creativity, extend existing programs at the expense of new and better ones, and perpetuate entrenched practices because "that's the way we've always done it." Combined with resource constraints, the result, as expressed by space consultant and educator David Webb, is that our space programs are designed "down to a cost instead of up to a standard."

Since the days when spaceflight made the transition from fiction to fact, society has continued to think of the Earth as a closed system. Meanwhile, space advocates have continued to envision our forays off the Earth in the framework of the Von Braun destination-driven paradigm. This has done nothing to change—and has probably reinforced—society's closed-Earth perception. The Moon, Mars, and other points of light in the night sky don't prompt most people to think about humanity's movement outward and the creation of new capabilities and new communities. Rather, if they consider human space efforts at all, they think of planetary journeys as elaborate geology field trips. You don't stay there, you just pick up your rock samples and go home.

Promoting the concept of Earth as an open system will require moving away from advocacy for particular destinations and methods of getting there, and replacing it with a reasoned, capabilities-driven approach. If we don't move in this direction, we risk expending considerable time and resources with no confidence that the resulting infrastructure elements can support our long-term strategic goals. If human missions target destinations without a well-crafted plan for ongoing productive activity there, we could repeat the post-Apollo scenario: the loss of a generation or more of progress, and the degradation of the government, private sector, and academic institutions that are essential for the exploration and development of space. The Apollo experience demonstrates that these unwelcome circumstances could come to pass *even if a destination-driven expedition succeeds.*

If humans are going to expand their economic sphere out into the solar system, as Marburger and others have suggested, then we'd better learn how to utilize the resources and unique environments we find there, preferably before we send large numbers of people across great distances, where the supply chain is far longer than in any situation previously experienced by our species. It will take decades to traverse this learning curve and overcome the technical hurdles, but that doesn't mean we should shrink from the challenge and withhold the investment. That would be a self-fulfilling prophecy, which is not the way to achieve self-actualization, or any other type of goal. On the contrary, we should get started as soon as possible. And we need to start with two primary rationales: economics (traditional space industries plus new endeavors like the harvesting of extraterrestrial resources and the processing of materials in microgravity) and survival (planetary stewardship and protection).

Putting economic and survival efforts front and center, rather than targeting increasingly distant destinations, may be viewed by some space advocates as dull or insufficiently visionary for humans who have ambitions of becoming a spacefaring species. I respectfully suggest that anybody who thinks this way should give it a bit more thought. A few decades of developing skills and building infrastructure focused on the Moon and near-Earth asteroids, and sending increasingly sophisticated robots to more distant locations, would enable sustainable movement outward and bring potentially critical benefits to Earth.

After we have the space infrastructure, experience, techniques for "living off the land," and wisdom to know what to do, other than basic science, at permanent human settlements in space, we will be able to choose a time for our first trips to other planets and be confident that we'll be successful and purposeful. When we go, we'll be ready to establish a sustainable, productive community, based on what we learned in Earth's environs. As individuals, we shouldn't be put off by the fact that we're unlikely to be around long enough to see the culmination of this process. After all, it's not about us, it's about the future of humanity. And in the meantime, we have plenty of work to do.

Chapter 8
The Century Perspective

To steer correctly, a system with inherent momentum needs to be looking ahead at least as far as its momentum can carry it. The longer it takes a boat to turn, the farther ahead its radar must see.—Donella Meadows, Dennis Meadows, and Jorgen Randers in *Limits to Growth: The 30-Year Update* (2004)

Changing our thinking

Earlier chapters discussed the strong drivers that cause strategic planning to be dominated by short-term thinking. However, there are some areas where incentives are shifting us to a more farsighted approach. An obvious example is America's economic future, along with closely related issues such as health care and the social safety net. Post-World War II Americans have tended to assume that each generation, on average, would do better than the last. For as long as that assumption proved true, we were generally confident that our kids would be alright financially, and that we'd all do fine in the long run through some combination of salary growth, employer-supported health insurance, investments, pensions, and Social Security. The biggest economic challenge for the nation's middle class throughout most of the post-war era has been maintaining adequate household income during downturns and increasing it during good times. Experience taught us that downturns were short-term fluctuations lasting from a few months to a year or so. Short-term planning seemed sufficient.

Developments during the past few years have undermined our confidence in this generally positive trend. The first decade of the new century saw the United States reaching deficit spending and a national debt far beyond anything we'd experienced or imagined before. At the state level, budgets fell dangerously into the red in most of the country. The housing boom went bust, taking down many homeowners and their financial institutions. Beacons of U.S. capitalism, including some that have been around for many generations like General Motors and Chrysler, found themselves in bankruptcy and looking for government bailouts. Nationwide, unemployment reached levels not seen in decades. Health care costs continued to rise and more than 45 million Americans had no health insurance. On the immediate horizon was the retirement of the baby boom generation, expected to throw Social Security into deficit spending by the middle-2010s. So much for relying on salary growth, health insurance benefits, investments, pensions, and Social Security.

All of this is happening at a time when many emerging nations have become serious competitors for resources and markets. There are a multitude of ways to illustrate the data on global population trends, but here's one I find staggering: by the early years of the 21st century, the combined populations of China and India exceeded the total world population of the 1940s. In just six decades, two Asian neighbors grew to numbers large enough to spread out across the planet—in the absence of all the rest of us—and still maintain sufficient population density to operate a modern industrial society.

The rest of the planet has grown as well, and will continue to do so. If official estimates prove true, the global population will reach nine and a half billion by mid-century—an increase of 40 percent from 2009. Demand for resources such as raw materials, food, water, and energy will go up by at least the same percentage, and in some cases—especially water and energy—at a much higher rate. Most of us know enough economics to recognize that increasing demand for finite resources means scarcity and higher prices. Traditional supply chains will be altered as countries that used to export raw materials decide to keep them for domestic use, or redirect their output to the highest bidder in fierce competitions that cast aside longstanding relationships. We're already seeing this happen with many rare minerals that are components of a wide range of electronic devices.

Uncertainties about national and global economic stability force us to think farther ahead and discard our assumption that fiscal problems will somehow work themselves out through economic growth and technological advancement, and everything will be fine. That could still happen, but we're starting to realize it's not the only possible outcome, and may not be the most likely one. The good news is that this realization is the first step toward individual and collective efforts to shape the future toward positive outcomes.

Environmental degradation and climate change are also motivators for long-term thinking. The literature and news coverage on these subjects almost always speaks in terms of consequences through the end of the century. The observed and projected effects are gradual so far, but we've become well aware that they're cumulative and could be difficult or impossible to reverse. Both mitigation and adaptation are important, and the earlier we start, the better. Space applications have already played a key role here, giving us some of the most important tools we have for detecting and analyzing the problems. The challenge for the rest of the century will be to use space capabilities to develop solutions as well.

The core resources

Our ability to accommodate a growing global population while improving the human condition and minimizing conflict depends on reliable, affordable supplies of resources that are equitably distributed. The core resources, upon which all activities depend, are energy and water. Everything we produce and consume requires these two resources, sometimes in surprisingly large amounts. Without these core resources, our agriculture, industrial processes, transportation, and information flows come to a stop. Even if other resources such as precious metals are abundant, inadequate supplies of energy and water—or more specifically, clean energy and fresh water—will be a showstopper, undermining the economy, the environment, and human health.

These core resources are already in short supply in many parts of the world, and as already indicated, this situation is poised to get worse, even before factoring in climate change. The U.S. Department of Energy, in its 2009 *International Energy Outlook*, estimated that the world's consumption of all types of energy will increase 44 percent between 2006 and 2030. The demand growth will not be evenly distributed. In the industrialized nations that are members of the Organization for Economic Cooperation and Development (OECD), the expected energy consumption increase during this period is 15 percent, while in the rest of the world's demand will go up 73 percent. (The OECD includes 30 industrialized countries in North America, Europe, and the Pacific Rim, not including China or India.)

When considering the Energy Department's projections, we need to keep in mind that they are based on existing trends in population and industrial growth. They don't consider the possibility of game-changing events (good or bad) or technological breakthroughs, just routine evolution in efficiency and capacity. The *Outlook* report recognized that changes in national and international laws and regulations on greenhouse gas emissions could restrict demand for fossil fuels, but this is not factored into the report's projections, resulting in a "business as usual" view of the future. Although renewable sources, particularly hydroelectric and wind power, are projected to grow significantly, fossil fuel use continues to grow as well and maintains its overwhelming dominance. Solar energy, according to the report, will still be uneconomical in 2030, except for niche applications.

The "business as usual" scenario could prove inaccurate due to technology breakthroughs, unforeseen discoveries of new resources, unexpected shortages (real or contrived), or international conflicts that disrupt global supplies. In any case, no single solution will provide the silver-bullet cure for our energy needs. Energy supplies will always come from a mix of sources. But that mix is not predestined if we choose to focus our resources on a different path. For example, we could choose to disprove the *Outlook* report's finding that solar energy won't be able to compete economically with other sources in the next two decades. Space technology for collecting and transmitting power can play a role in this effort.

Regarding the other core resource, the Earth has plenty of water, but only about 2.5 percent of it is fresh water, and much of that is locked up in snow and ice at high latitudes. The United States has a high per capita consumption of water, and the consumption rate in many emerging economies around the world is increasing at about twice the rate of population growth as they strive to improve living standards.

I don't intend to suggest that we'll find or fabricate water in space and haul it back to Earth to resupply our reservoirs, freshwater lakes, and underground aquifers. But space systems can help us better manage water resources through improved observation and by providing the energy needed to perform water treatment on a scale far larger than anything we've attempted to date.

A mid-century scenario

Many indicators have shown that the types of consumption and emissions behaviors typical in developed nations in the late 20th century will not be sustainable in the 21st century world. Energy will not be as cheap, resources will not be as abundant, yet population and industrial activity will continue to grow, or at least try to do so. The global environment appears to have reached the limit of its ability to be used as a sink for waste products in the form of pollution and greenhouse gases. The consequences of continuing these practices are not only unpleasant, they're self-destructive. Assuming that we don't have a collective death wish, we'll change the way we do things.

Some lifestyle changes are already evident around the country, such as increased participation in recycling programs, water use restrictions, and the use of higher-efficiency light bulbs and appliances. But this is just the modest beginning of a process that will transform America by mid-century. Here are some possible developments that may depend on or interact with space technology:

- Most of the nation's automobiles and light trucks will be converted to all-electric propulsion systems. They will have no tailpipe emissions, eliminating a major source of air pollution and greenhouse gases, and the electricity to power them can come from an array of sources, ending the monopoly of oil in the transportation sector.
- Following the lead of other countries, the U.S. will experience a renaissance in passenger rail, which also will be powered by electricity. Regional high-speed trains will expand their reach until they link up to connect the entire country. There may even be a latter-day version of the 1869 celebration at Promontory Summit, Utah that signaled the completion of the transcontinental railroad (perhaps on the bicentennial of that event), although it probably won't involve driving an engraved golden spike into a rail tie because the trains will be using magnetic levitation by that time.
- Electric utilities will be in the midst of a concerted effort, driven by both regulation and economics, to convert electricity production from fossil fuels to renewable sources. This will be helped by a number of new technologies and practices, including the widespread adoption of lightweight, efficient solar arrays in the construction and renovation of homes and businesses, easing the utilities' capacity requirements at peak hours.
- The number and capacity of water desalination plants will increase dramatically to compensate for underground aquifers that have shrunk or become brackish, diminished snow and glacial runoff from mountains, and increased demand for fresh water for agriculture, industry, and residential use.
- Continuous worldwide monitoring of Earth's ecosystems will be necessary to watch for symptoms of a sick biosphere, determine

the source of its affliction, and provide decision-makers with information needed to enable quick responses. Remote sensing from space will provide the majority of what the military would call "persistent surveillance" of a few dozen critical parameters affecting global environmental health.

These are not predictions of our fate, but possibilities that we can choose to pursue for a better future. If they seem ambitious, keep in mind the technological, economic, and societal evolution we've seen in similar timeframes, such as these examples:

- Regularly scheduled transoceanic passenger flights were initiated in the mid-1930s, in the midst of the Great Depression and just over three decades after the Wright Brothers' first flight.
- The U.S. Interstate Highway System included about 55,000 miles of roads by its 40th anniversary in 1996.
- Sputnik, the first artificial satellite, shocked the world in 1957 by simply beeping a radio signal from low Earth orbit, but by the 1990s hundreds of operational satellites in various Earth orbits were providing an assortment of vital services to governments, businesses, and households worldwide.
- The first general-purpose, mass-produced microcomputers appeared in the mid-1970s. Thirty years later, desktop, laptop, and handheld computers were ubiquitous in homes, schools, and workplaces. Their capabilities increased by orders of magnitude during that time while their prices went down dramatically. Generations of young Americans can't imagine life without them, or without the link to the world provided by the Internet.

Science fiction writer Vernor Vinge, inventor and futurist Ray Kurzweil, and others have argued that knowledge and technological advances are increasing *exponentially*, offering examples and justifying why they believe this is situation will persist. I suspect that depending on what metrics are used for "knowledge" and "advances," they are correct in at least some areas of human development, for at least some portion of human history. Progress is uneven, as evidenced by the fact that the public policy evolution, which is needed to help society cope with these advances, lags far behind the science and technology. Those of us in the policy community would be elated if we could make continuous *linear* progress. Nonetheless, belief in exponential growth across a wide range of human endeavors is not required to recognize the potential for great change within a generation or two, and the possibility that space will be a critical component of it.

Enabling the future with power from space

All of the energy that powers our society today, with the exception of nuclear and geothermal, comes from space. Fossil fuels started out as plants and animals that were dependent on the Sun, so they wouldn't be here without solar energy deposited on Earth eons ago. Today's solar energy collectors have an obvious direct link to the Sun, which also drives the wind and the hydrological cycle, and is essential for biomass growth. The Moon and the Sun both affect ocean tides, which may someday become a substantial source of energy. Taking all of this into consideration, deriving energy from space shouldn't seem like anything out of the ordinary.

Princeton physicist Freeman Dyson is one of the early theoreticians associated with the idea of deploying a network of solar collectors in a shell around the Sun. When he wrote on this topic in the 1950s, he felt that a growing technological civilization would get to the point where it would need to capture as much output as possible from its sun. Science fiction writers extended this idea to an extreme: use all the matter from the planets and asteroids orbiting a star to enclose the star in a sphere that would capture all of the energy. We don't need to go that far. Peter Glaser's solar power satellite (SPS) concept for geosynchronous orbit, introduced in 1968 and described in Chapter 5, is an appropriate undertaking for our time.

As noted earlier, the most serious objection to SPS offered by opponents has been the cost. Such objections tend to ignore the fact that investments in developing and deploying SPS would likely be spread over the next three to four decades, during which time *trillions* of dollars in new electricity generating capacity will be required on Earth. So far, U.S. investment in SPS has been paltry and inconsistent. In the 40 years that followed Glaser's introduction of the SPS idea, the U.S. government's on-again, off-again research funding in this area totaled a mere $80 million, according to a study by the Pentagon's National Security Space Office. Most of that came in the late 1970s and late 1990s. A much more ambitious program could be conducted today for a level of government investment that would not involve an unreasonable amount of money, and would probably find some partners in the private sector willing to risk their capital.

SPS research could thrive on a fraction of the funding that's been lavished on other energy interests. Research and tax subsidies to the fossil fuel and nuclear industries since the mid-20th century have been consistent and massive, and fusion energy research has received an average of around $400 million per year for the past 50 years. Fusion has long been chided as the energy solution that's perpetually 30 years in the future, and the reality is even less promising given the urgent need to develop alternatives to fossil fuels. Larry Grisham of Princeton University's Plasma Physics Laboratory wrote in his chapter of the 2008 book *Future Energy* that he expects

> ... the construction of a demonstration power plant in the middle of the century, which could lead to the deployment of commercial fusion power plants in the second half of this century and, if they prove sufficiently reliable, a very large contribution to the world's energy needs in the next century ...

Not a very hopeful forecast for a society seeking solutions in the 21st century. Despite this, fusion research has received about a quarter-billion dollars a year in federal funding for the past decade. Stated another way, fusion's *annual* federal support in recent years has been more than three times the total amount SPS has received *in 40 years*. If we were to start today to consistently provide SPS with a level of funding equivalent to fusion, I believe SPS would be up and running first, making a significant contribution to energy needs many years—perhaps decades—before fusion is ready to do the same.

The commitment to SPS should not be judged in isolation from other space efforts. If we plan to pursue related technologies anyway—larger launch vehicles, advanced space robotics, assembly of large space structures, exploration of the Moon and asteroids—SPS development will benefit directly from all of this investment in new capabilities, including the use of extraterrestrial materials for construction. The space efforts of other nations will provide a boost as well, since many countries are working along similar lines and are interested in investigating the possibilities of SPS.

To stimulate government investment in SPS development, there needs to be a reasonable expectation of societal benefits that would most likely not be realized, or would be substantially delayed, without government assistance. Feeding the electric utility grid may or may not be enough by itself, but the combination of multiple potential benefits make it worthwhile to invest in SPS along with other options for producing large amounts of power while avoiding greenhouse gas emissions.

Commuter and intercity rail transportation is a power-hungry application that will become increasingly important in our mid-century scenario. Commuter trains and high-speed rail powered by electricity will be essential for relieving travel congestion within and between cities while also reducing pollution and carbon emissions. Intercity high-speed rail doesn't exist yet in the United States, but it can be found in Japan, Taiwan, several countries in Europe, and elsewhere, so the U.S. has some catching up to do. One of the great things about converting to electric transportation is that electricity can be generated in a variety of ways using an assortment of resources.

The bad news is that today, most of the energy for our commuter trains and buses comes from fossil fuels. Despite this, using public transportation rather than individual gasoline-powered vehicles yields a net reduction in greenhouse gases. A September 2007 report by Science Applications International Corporation (SAIC), using data from 2005, determined that carbon dioxide emissions in the U.S. that year were reduced by 6.9 million metric tons, and other greenhouse gases by 400,000 metric tons, due to use of public transportation instead of personal vehicles. However, the public transportation systems themselves produced 12.3 million metric tons of carbon dioxide, so there's still plenty of room for improvement.

Water treatment is also a big consumer of power, and the situation will get more demanding as population grows and the requirement for desalination capacity increases dramatically. Worldwide, there are approximately 15,000 desalination plants of various sizes, but unfortunately most of them get their energy from fossil fuels. The energy efficiency of the pumping and filtration process has improved dramatically over the past 40 years, but according to a study by the U.S. National Research Council, it's within about 15 percent of the theoretical maximum efficiency, so little improvement can be expected. It can take as much as 14 kilowatts of energy to produce 1000 gallons of desalinated seawater.

The United States consumes about 400 billion gallons of water per day according to the U.S. Geological Survey, and less than 0.4 percent of this comes from desalination plants. As of 2006, worldwide desalination capacity was about 10 billion gallons a day, which means that if the U.S. alone could match that output, it would accommodate only 2.5 percent of the nation's water needs. If the country wanted to get, for example, 25 percent of its water supply from desalination, it would have to operate the equivalent of 500 plants, each producing 200 million gallons per day. (The largest plant in the world, in Jubail, Saudi Arabia, produces 211 million gallons per day.) A challenge at least as big as building hundreds of large desalination plants would be providing each one with a dedicated gigawatt-class power station.

Obviously, running all those power stations on fossil fuels would be unsound policy. In addition to the greenhouse gas emissions, the increased demand for fuel would be enough to disrupt energy prices. Although coal is the most abundant domestic fossil fuel resource, so-called "clean coal" may never provide an adequate solution. Outside of the research community attempting to develop usable technologies, it exists primarily as a marketing slogan for the coal industry and a lobbying campaign for federal research dollars. Routine commercial carbon capture and sequestration is many years and tens of billions of dollars away, and even if it can be proven to work, the result may never justify the investment. Energy consumers, meanwhile, will lose decades waiting for the new technology to be perfected and widely adopted, and through their tax dollars they will be subsidizing research that artificially props up the coal industry at the expense of technologies that would enable a transition to sustainable energy resources.

The advantage coal has today is that its large domestic supply makes it cheaper than alternative sources for large-scale power production. But new plants with carbon capture technology, and old plants that are retrofitted at great expense, will lose that advantage because it will cost significantly more to produce power. It's too early to pin this down precisely, but I've seen estimates of electricity cost increases ranging from 15 percent to more than 100 percent, depending on the technology and the type of coal used. And that's for new power plants. Retrofitting older plants will add at least another 30 percent.

The added production costs come from separation and compression of the emission products, transportation to the storage site, and injection into geological storage. Researchers developing the techniques estimate that the energy penalty of the compression stage alone is about 10 percent of a coal plant's output, and operations and maintenance costs must be added to that. The costs of the transportation and injection stages are dependent on the distance to the storage site (which could be considerable), and the depth and type of geologic storage. Even if the storage facilities could be guaranteed never to leak their contents into the atmosphere (thus defeating their purpose), the investment in time and resources on this approach rather than more sustainable ones doesn't make sense from a societal perspective. If the coal industry really believes in this, it should continue the research with its own resources and attempt to prove its ability to clean up its act and still compete successfully.

Biofuels aren't a good solution either. They're aimed primarily at providing liquid fuels for the transportation sector, which is expected to be mostly converted to electric propulsion in our mid-century scenario, although there may still be a smaller market consisting of large vehicles that would benefit from some form of biodiesel. However, it's already been demonstrated that conversion of agricultural land from food to fuel production is the wrong approach. This has caused food price hikes around the world as U.S. corn growers have switched their crops to fuel markets.

Biofuels wouldn't do more than make a token contribution to the resource demands of power plants. Conservation groups have calculated that as a gasoline substitute, replacing just five percent of U.S. consumption would require inordinate amounts of farmland (the equivalent land area of Oregon and Idaho for corn ethanol, the equivalent of New York and Connecticut for cellulosic ethanol). In any case, production of biofuels uses so much water—a problem it shares with coal mining and processing—that it would be at odds with the purpose of desalination plants, or simply exacerbate the water problem in another location. The refining process requires four gallons of water for each gallon of fuel produced, and that doesn't include the water used in the field to grow the feedstock, which will vary by location.

Nuclear power may provide an interim solution, but it brings its own set of problems, including security and waste disposal issues. Another consideration is the cost of building the power plant, which in all probability will exceed the cost of building the desalination plant.

Terrestrial renewable energy sources are better, but not perfect, solutions. Hydropower, geothermal, and ocean current resources are location dependent, and to some extent so is ground-based solar because its output is affected by cloud cover and latitude. Similarly, wind is available everywhere, but varies by location and weather conditions. A desalination plant in Perth, Australia produces more than

26 million gallons of water per day using 23 megawatts of electricity from wind, but this would be considered a relatively small plant in our mid-century scenario. The filtration process is most efficient when it runs continuously, but neither wind nor ground-based solar can guarantee a steady supply of energy. In contrast, SPS will be designed as a continuous power source, except for brief eclipse periods that total only a few hours a year. When the SPS system reaches maturity, it will deliver higher power levels with less need for energy storage than ground-based solar or wind energy.

Depending on the typical size of the desalination plants and the SPS platforms serving them, the output of a single platform may serve multiple plants. As communications satellites do today, the power platforms could direct multiple, reconfigurable spot beams at numerous receiving antennas on the ground.

In addition to the electrification of the transportation sector and growth in power needs for water treatment, there are security applications for SPS as well. Numerous studies have found that dependence on fossil fuels undermines national security at both the micro and macro levels. At the micro level, over-reliance on oil is a costly burden for the military that undermines combat effectiveness and endangers lives. For overseas deployments, the full cost of a gallon of gasoline is at least $15 to $42, and by some estimates as high as $400, when burdened with all transportation and security costs. As much as 70 percent of the tonnage of supplies delivered to forces in Iraq has been petroleum products, a substantial percentage of which is used to fuel generators for electric power that may someday be obtainable from an SPS constellation. At the macro level, U.S. dependence on imported fossil fuels weakens international leverage and economic stability while entangling the country with unstable or hostile regimes. Energy shortages and price hikes will cause problems for other countries as well, contributing another serious impediment that could turn weak or unstable regimes into failed states.

The study by the Pentagon's National Security Space Office, released in October 2007, specifically addressed the role SPS could play in alleviating these concerns. The report found that:

- SPS would enhance the nation's energy security.
- The Department of Defense could be an active user of services provided by an SPS system, both at its fixed bases and as a supplement to the power needs of deployed forces.

- By providing an alternative to fossil fuels and nuclear power, SPS would mitigate resource scarcity, climate change, and the motivation for more countries to develop nuclear capabilities, all of which may help to avert future conflict.
- A strategy to deploy SPS would stimulate efforts to develop cheaper, more reliable Earth-to-orbit launch systems, which would also benefit other space endeavors.

A variety of terrestrial alternatives for accommodating the enormous growth in energy demand are available and will undoubtedly contribute to the nation's and the world's energy mix. But as we assess the relative costs—including environmental costs—in the years ahead, it will become evident that Earth-based attempts to meet energy demand have price tags just as daunting as space-based attempts to boldly go where no electric utility has gone before.

The role of the space community from a century perspective

It's time for space technology to save the planet—again. It did this once already, during the Cold War. Most people may believe that the triad of offensive nuclear weapons based on land, sea, and air was responsible for keeping the peace, but that's only part of the story. Our nuclear capability gave us both deterrence and the potential for doomsday. The success of the former and avoidance of the latter owes substantial credit to satellites, primarily those used for intelligence and communications. They kept leaders on both sides informed and connected in ways that prevented the hair trigger from being pulled. And by the 1980s, satellites were among the best tools available for reaching out to the people of the Soviet Union, and developing nations as well, to encourage them to connect with the rest of the world in beneficial ways. Global nuclear war was averted, the Soviet empire was dismantled, and much of the credit goes to space technology, which provided desperately needed capabilities at the right time.

Space technology can be called upon again to come to the world's rescue in the very different circumstances of the 21st century. As in the Cold War, it's not the sole solution, but it can be the key element that puts success within reach. As before, it will bring us intelligence, which it already has done in abundance in the form of scientific knowledge and ongoing environmental monitoring. It will also continue providing communications to foster integration of international research and innovation efforts. These are capabilities that space systems have been building on for decades. There will also be new developments that will lead the way to the energy and material wealth of the solar system. But how do we get there from here?

Collectively, the space community, which has a tendency to be too insular, must open its intellectual aperture wider than the space visionaries of the past and seek to encompass the highest-salience global challenges of our era. This new generation of space futurism requires going beyond the disciplines typically associated with space science and engineering and becoming more than a technologist's dream factory. A multidisciplinary approach is essential, but this doesn't simply mean applying multiple types of engineering. Political, economic, and other societal developments must be considered in parallel, including topics such as geopolitics and international relations, economic and technological globalization, and global threats, both natural and man-made. Such an approach to space development will attract a wider public and private sector constituency, bring greater stability to the allocation of resources, and once again attract the brightest and most energetic minds to space-related fields. It also will redound to the benefit of space exploration ambitions, since exploration and development go hand-in-hand.

The OECD, the organization of industrialized countries mentioned earlier, released a two-part study in 2004 and 2005 that provides a clue to the evolving nature of national space agencies and the rest of the space community. Addressing the future of space applications through the 2030s, the study sought "to understand how OECD countries may reap the benefits of civil and commercial space applications for society at large" using scenarios based on "the interaction of three main drivers of social change: geopolitical, economic, and environmental." The recommendations lean heavily on the establishment of sustainable space infrastructure, with emphasis on maximizing private-sector involvement in developing and operating the infrastructure while governments set the stage with open markets, international standards, and laws that are friendly to space businesses. Overall, the OECD recommendations align well with the idea that space development should continue to play a significant role—far beyond just communications—in shaping globalization's evolution and keeping it focused on societal needs. At least for the next three decades, the study foresaw great hope and promise for applied space research and development, with government space agencies continuing to spearhead basic research.

This brings us back to the quote from visionary Dandridge Cole offered in Chapter 2: "We could fill books with problems of fundamental importance to the human race which can be solved *only* by spaceflight, *more easily* by spaceflight, or *more probably* by spaceflight." Cole did not finish the job of filling the books that he alluded to, though he might have tried had he not died of a heart attack at the age of 44. The task remains for us to complete.

Chapter 9
Commitment to the Future

Creativity and imagination are necessities, not luxuries.
—Robert A. Cassanova, former director, NASA Institute for Advanced Concepts

Tomorrow is our permanent address.—Marshall McLuhan, media analyst

At the beginning of the space age, everyone in the embryonic space community had to be a futurist. Missions had to be proposed, based on requirements that would remain valid for many years; space systems and ground infrastructure had to be designed and built. The space community is much larger today, and has settled into ongoing operations that provide the communications, Earth observation, and navigation services discussed in previous chapters. But although space applications are integrated into the fabric of society, and many other nations have developed their own space capabilities, forward-thinking is no less important to the United States today than it was at the dawn of the space age.

Space was the cutting-edge, fast-moving pursuit of the 1960s, but today it's viewed in the U.S. as one of the slower-paced sectors of high technology, surpassed in speed and agility by information technology, biomedical advances, and other disciplines. This perception, fed by widely reported problems in controlling costs, meeting schedules, maintaining quality, and retaining global market share in space technology, evokes images of a community that has lost its edge and its relevance. This is occurring at a time when the geopolitical and economic environment—which directly affects the missions, priorities, and resources of the space community—is evolving rapidly. In order for the space community to remain relevant, it must be vigilant of societal and technical trends, not only in space-related fields but in other key areas that directly and indirectly use or affect its work. A sustained effort is required to integrate these trends and formulate scenarios extending well into the future.

In the United States today, there is no strategic imperative for space exploration and development in general, or for human spaceflight in particular. All too often, space officials and advocates answer the questions "Why" and "Why now?" with vague platitudes like "It's our destiny" or "It will inspire our children." Even in the media event at NASA Headquarters on January 14, 2004 in which President Bush announced his exploration initiative, he called it "a journey, not a race," emphasized human destiny and the desire to explore, and stopped short of identifying it as a national imperative. Since that time, the lack of urgency has been reinforced by a lack of resources, in sharp contrast to the NASA budget of the early 1960s, which increased dramatically to accommodate Apollo. As a result, expressions of disinterest and disapproval have been common from an electorate more concerned about overseas military conflict, record federal deficits, health care, and the economy. Maybe space projects would garner more attention and support if the answer to the "Why spaceflight?" question is not "because it is there" but rather "because that's where the resources are, and because it will save our butts." Of course, that would require more than an eight-second sound bite to explain.

A 2003 RAND study on long-term policy planning made a general observation that rings true for the majority of space planning since Apollo: "[M]ost futuristic narratives are created with the aim of commenting on and shaping the present rather than supplying an accurate roadmap for what is to come." This has been particularly noticeable in recent years. Rather than tackling the big-picture issues like how to use space capabilities to improve the human condition and expand horizons of knowledge and experience, U.S. civil space policy has concentrated far more on fixing or terminating troubled programs, preserving the nation's declining space science and engineering expertise, preventing a gap in human spaceflight, and locking in our next-generation launch architecture before we have a good grasp of what we're going to do with it. All of these are short-term concerns in the grand scheme of things, and would be substantially alleviated in an environment where the overarching purpose was known and accepted, and the steps to get there were mapped out sufficiently to clarify requirements, win consistent political support, and convince the prospective workforce that a career in the space community could promise a rewarding future.

There is a place for space in national and global affairs, if the space community can become less inward-looking and national decision-makers can take a longer view that incorporates space capabilities beyond those in current use. As physicist and visionary Gerard K. O'Neill noted in his book *2081*:

> The long-term health of a nation is probably shown most clearly by the time scale of the programs it undertakes. The willingness to commit to ventures of many years' duration, with potential very large returns, is the hallmark of a nation confident of its own future. The fear of any commitment beyond one or two years is the symptom of disease, signaling a fundamentally hopeless view of the future and the intention to cut the losses and get out of the game.

Numerous challenges face our planet in the coming decades that will require us to "commit to ventures of many years' duration." Many of them, including high-visibility concerns such as environmental degradation and climate change, are amenable to the application of space technology. As these problems move higher on the public agenda, the nation and the world will look to space technology for information and solutions, and the space community should be ready to answer that call. In fact, our efforts should allow us to be prepared far in advance so that plans and programs are already in place when the call comes.

Making the future a better place through space applications will require thinking differently than we did in the Cold War. In the midst of that era, Arthur C. Clarke offered a caution in his 1963 book *Profiles of the Future* that is still valid today:

> [T]here is rather more to space exploration than shooting men into orbit, or taking photos of the far side of the Moon. These are merely the trivial preliminaries to the age of discovery that is now about to dawn. Though that age will provide the necessary ingredients for a renaissance, we cannot be sure that one will follow. The present situation has no exact parallel in the history of mankind; the past can provide hints, but no firm guidance.

Looking back at the time when Clarke wrote these words, aficionados of spaceflight history often contemplate whether the space program was the result of natural human evolution or a Cold War fluke. This is an interesting and informative academic exercise, but it teaches us more about how *not* to do long-term strategic planning than about how it should be done. Our first rule needs to be: Start from where we are right now.

The current era is breaking many of the rules of the Cold War, particularly with regard to the spread of information and the balancing of international cooperation and competition. Knowledge moves faster than ever, expertise is proliferating, and technology transfer must be treated not simply as an outflow that robs us but as an inflow that enriches us. The knowledge-based activity that space efforts thrive on must draw from around the world. Attachment to closed or restrictive environments, beyond what is actually necessary, will drive talent elsewhere.

As we purge our Cold War mindset and prepare for the future, we also need to reconsider the long-held belief that our civil space program should be designed around flagship missions in human spaceflight. The effect has been to set up false dichotomies: humans *or* robots; exploration *or* development; low Earth orbit *or* the Moon and beyond. The resulting arguments have produced plenty of heat but little light. We should be planning a space program of exploration *and* development, where the choices of long-term goals and the milestones to get us there determine the balance between humans and robots, and between activities in near-Earth space and journeys elsewhere in the solar system.

As humans and robots each take on the roles for which they're best suited, we needn't worry that astronauts will have nothing to do. There will be plenty of work for people in the near-to-medium term in low Earth orbit, on the Moon, and at various points in between. That's where we'll construct the advanced space systems that bring benefits to Earth, while simultaneously building the technologies, skills, confidence, and wisdom to venture beyond the Moon and actually know why we're doing it.

Some may still fret for the popularity of astronauts, and for public support for space in general, if people get busy in the Earth-Moon system and are in no hurry to set off to increasingly distant points in the solar system. But I don't expect that astronauts in the decades ahead will be disparaged because they "only" went to the Moon, especially if they pave the way for others to go there. Spacefarers will generate just as much excitement as they always have when they show up to give a talk at your kid's school, especially if they talk about all the productive things they're doing that set the stage for a better tomorrow. As the Earth-Moon system develops into an industrial park for microgravity manufacturing and the harvesting of lunar materials and solar energy, there will be more astronauts with more to say. Maybe by then we'll think of them as space workers, with jobs that young people can aspire to, rather than an elite corps that only a select few can join.

There's no doubt that the folks back home identify with human adventurers who go to exotic and dangerous places, but would they really support a succession of touch-and-go missions to Mars, and then the asteroids, and then the moons of Jupiter, and so on? If we embark prematurely on such missions, touch-and-go is all we'll be able to accomplish, and we won't go back again for a long time, just as it happened with Apollo. Would the folks back home feel that the bragging rights were worth the risk and cost, especially knowing that the missions could be performed far more efficiently (and in some cases, already have been) by robots?

A favorite retort of human spaceflight advocates to such a query has been, "They don't give ticker-tape parades to robots!" True, but they don't give ticker-tape parades to astronauts anymore, and NASA's robots have websites that attract millions of hits. Never underestimate robot charisma.

Most people born in the U.S. in the post-Apollo era grew up in a world with computers everywhere—at home, at school, and in the workplace. General-purpose computer workstations are the most obvious examples, plus there are a multitude of special-purpose computers embedded in toys, games, and household devices that we take for granted. We've also become accustomed to the entertainment industry's depiction of robots as smart and capable. Personal identification with machines comes naturally to post-Apollo generations.

Starting in the 1990s, NASA began to explore Mars with rovers instead of stationary landers. The mobility has been a boon to scientific investigations, but also may have profoundly changed the public's perception of the missions in unexpected ways, especially among the younger and more technology-savvy population. A robot that moves around the surface of another planet generates a whole new level of interest—and expectations for the future—compared to a robot that simply swivels a camera platform from a fixed position.

Evolving public perceptions of robotic systems indicate that NASA should emphasize not only the complementary nature of humans and robots on space missions, but also the centrality of robotics to the exploration and development architecture. The public is eager to see technological advances coming from their tax dollars, and can appreciate that the use of robots contributes to the safe and efficient performance of tasks, saving money compared to human systems in many applications. Technophiles, especially younger ones, also have expectations that robots eventually will proliferate the way computers have done, and space technology investment can accelerate this and give the U.S. a competitive advantage in the world market.

Many people in the space business believe that young children naturally become space enthusiasts, but then lose interest somewhere between junior high and college for reasons which remain unclear. This may be a faulty assumption—a substantial percentage of American children may not be space enthusiasts, and the onset of puberty may have no effect on their interest or attentiveness to the subject. It's hard to be sure, because polling organizations don't survey age groups under 18, and I'm not aware of any sociological study that has specifically addressed this question. At the same time, it seems clear that upcoming generations are increasingly comfortable with technology. Hopefully, their curiosity will prompt questions like: How do they make all those maps on the Weather Channel? How does the satellite TV signal get to my house? How does my GPS gadget figure out my location? Eventually, these can evolve into more sophisticated questions, such as: How can Earth observation capabilities be improved to give us even better information about weather, climate, and pollution? If a satellite can beam TV signals to multiple receivers on the ground, can it also beam power? What else can be done to take advantage of the characteristics and resources available in the space environment?

This kind of creative thinking needs to flourish, and I believe it can if we articulate goals and objectives that advance the exploration and development of space while remaining relevant to national needs and aspirations. That means thinking long-term, but not with a fixation on destinations. The search for capabilities and knowledge should determine the destinations, the pace, and whether humans get to go. It should not be the other way around, with destinations driving all of our other choices.

Recognizing that the infrastructure elements and other advanced systems we build have life cycles of 30 or 40 years, and sometimes longer, we need to make a commitment to the long-term future, both at the national level and at the personal level. This is true of society's efforts generally, not just the space enterprise.

Within the space community, perhaps the best recent statement of how we should direct our thinking and planning came from a panel formed by the U.S. National Research Council's Space Studies Board and charged with examining the rationale and goals of the U.S. civil space program. (The group was commonly referred to as the Lyles Committee, after its chairman Lester Lyles, a retired Air Force general with considerable space-related experience.) Their 2009 report didn't receive the attention it deserved, perhaps because it was released during the busy first year of the Obama administration when countless other reports recommending change were emerging, including a significant number that addressed space. A sample of the Lyles Committee's words of wisdom are worth repeating here.

> What will the next 50 years bring? . . . Many of the pressing problems that now require our best efforts to understand and resolve—from terrorism to climate change to demand for energy—are also global in nature and must be addressed through mutual worldwide action. In the judgment of the Committee on the Rationale and Goals of the U.S. Civil Space Program, the ability to operate from, through, and in space will be a key component of potential solutions to 21st century challenges. [The Committee appears to agree with Dandridge Cole, as quoted at the end of the previous chapter.]
>
> The national priorities that informed the committee's thinking include ensuring national security, providing clean and affordable energy, protecting the environment now and for future generations, educating an engaged citizenry and a capable workforce for the 21st century, sustaining global economic competitiveness, and working internationally to build a safer, more sustainable world.
>
> The committee's overall conclusion is that a preeminent U.S. civil space program with strengths and capabilities aligned for tackling widely acknowledged national challenges—environmental, economic, and strategic—will continue to make major contributions to the nation's welfare . . . The time has come to reassess, and in some cases reinvent, the institutions, workforce, infrastructure, and technology base for U.S. space activities.

The Lyles Committee's first recommendation—which drives the ones that follow, and is therefore the most important—is "*Addressing national imperatives*. Emphasis should be placed on aligning space program capabilities with current high-priority national imperatives, including those where space is not traditionally considered." This is good advice to ensure that our space efforts will remain a critical element of a national commitment to the future.

Worthwhile national goals in space development—important new capabilities, access to vast new resources, sustainable ways of generating wealth, accumulation of prodigious amounts of scientific knowledge—will take decades to achieve and will involve high cost and high risk. This challenge is also burdened with expectations of short-term gratification, in contrast to great human achievements of earlier eras like the cathedrals of Europe, the Egyptian pyramids, or the Great Wall of China, which were expected to be multi-generational projects. Cultivating a long-term perspective is the only way to produce valid long-range goals for our space efforts, and maintain the support networks needed to sustain implementation for the long haul. NASA, the nation, and our partners in these endeavors can't afford the false starts that short-term thinking engenders, especially since the welfare of the planet may depend on our decisions and actions right now.

Individual commitment to the future in many ways is more difficult to achieve than a societal commitment. Unlike societies, individuals don't expect to last for hundreds of years, so their thoughts of the long-term future may not extend beyond the financial legacy they want to leave to their immediate descendants. People who are in dire straits in the present don't even have the luxury of thinking that far ahead. The path to Maslow's self-actualization is unique to each of us, and some face more challenges than others, often putting the future on hold.

For space enthusiasts who grew up in the 20th century and remember looking ahead to the early 21st century as a future where humanity would be a multi-planet species engaging in ubiquitous spaceflight activities, there exists much cynicism because the reality is disappointing. In addition to the successes and frenetic pace of the Apollo program, they remember visionaries like Willy Ley, a space author who wrote for both children and adults, who said in his 1964 book *Beyond the Solar System* that he believed a manned mission to the nearby star Alpha Centauri "will be made at a time when people now alive (though very young) will be able to watch the take-off on television—say, half a century from now." Today, we know that it will be a very long time after 2014 when humans set out for another star. But instead of grousing about it and looking for someone to blame, we should expend more of our energies on making sure we continuously move forward, and less on worrying about whether all of our spaceflight dreams will be fulfilled within our lifetime. This should not be viewed as resigning oneself to meager achievements or failure. Rather, it's a recognition that we each play a small but potentially important part in an endeavor that is spectacularly large and crosses many generations. Pretty exciting, when you think about it.

For myself, I don't have any concerns about whether or not I get to travel beyond Earth. My personal and professional interests are not about Me in Space, but they are fulfilling nonetheless. I devote my efforts to making sure the people who come after me can maximize the benefits they get from space, and go there themselves if they want to. Some readers may find the suggestions in this book overly ambitious because they endorse large-scale space industrialization. Other readers may believe they don't go far enough because there's no call for the rapid spread of humanity around the solar system. If that puts me in a middle-of-the-road position, so be it—as long as that road is steadily going somewhere.

The more I study and work in this area, the more I am convinced that the drive to explore and develop space is not merely a Cold War anomaly, not a fad that we'll eventually abandon, and not just the province of rich countries representing a minority of the planet's population. Our species is at the earliest stages of a type of transition that has recurred throughout our history, each time changing us forever by pushing back physical limits and putting new sources of materials and energy within our reach. Finding the appropriate way to tackle the challenge of space is critical to providing societal benefits to Earth and to determining which nations and industrial sectors will share in this enterprise. Along the way, there's a lot of work to do, and there are no guarantees.

British scientist James Lovelock is known for developing the Gaia hypothesis, in which Earth is portrayed as a living organism. He is also known for being very pessimistic about humanity's future on our deteriorating planet. In an April 2009 interview, he stated his belief that it's very unlikely people will be going into space at the end of this century. According to Lovelock, "We'll be far too busy surviving to have the time or the money or the energy to do it, and it won't be considered that important then." This is certainly within the realm of possibility, but it need not be our fate. Those of us who want a more positive outcome that includes a spacefaring civilization had better get busy and start thinking and planning for the long term so we can prove Lovelock wrong. It's our choice.

Bibliography

Chapter 1: Cruising to Utopia—or Not

Bush, Vannevar, "Science—The Endless Frontier," report to the president by the director of the Office of Scientific Research and Development, July 1945 (*http://www.nsf.gov/od/lpa/nsf50/vbush1945.htm*).

Byerly, R. & R. Pielke, "The Changing Ecology of United States Science," *Science, 269*(5230), 1995, pp. 1531-1532.

Frieden, Jeffry A., *Global Capitalism: Its Fall and Rise in the Twentieth Century* (New York: W.W. Norton & Company, 2006).

Friedman, Thomas L. *The Lexus and the Olive Tree* (New York: Farrar, Straus, & Giroux, 1999).

Friedman, Thomas L. *The World is Flat: A Brief History of the Twenty-First Century* (New York: Farrar, Straus, & Giroux, 2005)

International Forum on Globalization (*http://www.ifg.org/*). The IFG claims to represent over 60 organizations in 25 countries and identifies itself as "an alliance of sixty leading activists, scholars, economists, researchers, and writers formed to stimulate new thinking, joint activity, and public education in response to economic globalization."

Homer-Dixon, Thomas. *The Upside of Down: Catastrophe, Creativity, and the Renewal of Civilization* (Washington: Island Press, 2006).

Kennedy, Paul. *Preparing for the Twenty-First Century* (New York: Random House, 1993).

Mack, Pamela E. *Viewing the Earth: The Social Construction of the Landsat Satellite System* (Cambridge: MIT Press, 1990).

Meadows, Donella, Dennis Meadows, Jorgen Randers, William Behrens III, *The Limits to Growth: A Report to The Club of Rome*, 1972 (executive summary at *http://www.clubofrome.org/docs/limits.rtf*).

Meadows, Donella, Dennis Meadows, & Jorgen Randers. *Limits to Growth: The 30-Year Update* (White River Junction, Vermont: Chelsea Green Publishing Company, 2004).

Moltz, James Clay. *The Politics of Space Security: Strategic Restraint and the Pursuit of National Interests* (Stanford, CA: Stanford University Press, 2008).

National Intelligence Council, *Global Trends 2025: A Transformed World*, NIC 2008-003, November 2008 (*www.dni.gov/nic/NIC_2025_project.html*).

Reich, Robert B. *The Future of Success* (New York: Vintage Books, 2000).

Shiva, Vandana, "The Polarised World of Globalisation," International Forum on Globalization, May 27, 2005 (*http://www.zmag.org/sustainers/content/2005-05/27shiva.cfm*).

Smil, Vaclav. *Global Catastrophes and Trends: The Next Fifty Years* (Boston: MIT Press, 2008).

Stiglitz, Joseph E. *Globalization and Its Discontents* (New York: W.W. Norton & Company, 2003).

Chapter 2: Searching for a Vision of the Future

Augustine, Norman R., "Report of the Advisory Committee on the Future of the U.S. Space Program," NASA, December 1990.

(Augustine Committee) Review of Human Spaceflight Plans Committee, "Seeking a Human Spaceflight Program Worthy of a Great Nation," October 22, 2009 (*http://www.nasa.gov/pdf/396093main_HSF_Cmte_FinalReport.pdf*).

Bush, George H.W., "Remarks by the President at 20th Anniversary of Apollo Moon Landing," July 20, 1989.

Bush, George W., "U.S. Space Exploration Policy," National Security Presidential Directive (NSPD) 31, January 14, 2004.

Clarke, Arthur C., "Extraterrestrial Relays," *Wireless World*, October 1945.

Clarke, Arthur C. *Profiles of the Future* (New York: Harper & Row Publishers, 1963).

Clarke, Arthur C. *The Promise of Space* (New York: Harper & Row Publishers, 1968).

Cole, Dandridge & Donald Cox. *Islands in Space: The Challenge of the Planetoids* (Philadelphia: Chilton Company, 1964).

Cole, Dandridge. *Beyond Tomorrow: The Next 50 Years in Space* (Amherst, Wisconsin: Amherst Press, 1965).

Cornish, Edward. *Futuring: The Exploration of the Future* (Bethesda, Maryland: World Future Society, 2004).

Disney DVD, "Tomorrowland: Disney in Space and Beyond," released May 18, 2004. Includes "Man in Space," originally aired March 9, 1955; "Man and the Moon," originally aired December 28, 1955; and "Mars and Beyond," originally aired December 4, 1957.

Kahn, Herman & Anthony J. Wiener. *The Year 2000: A Framework for Speculation on the Next Thirty-Three Years* (New York: Macmillan Company, 1967).

Kahn, Herman, el al. *The Next 200 Years: A Scenario for America and the World* (New York: William Morrow & Co., 1976).

Kennedy, John F., speech to a joint session of Congress on "Urgent National Needs," May 25, 1961 (*http://www.jfklibrary.org/Historical+Resources/Archives/Reference+Desk/Speeches/JFK/003POF03NationalNeeds05251961.htm*).

Launius, Roger, "Looking Backward/Looking Forward: Space Flight at the Turn of the New Millennium," *Astropolitics*, Vol. 1, No. 2, Autumn 2003, pp. 64-74.

NASA, "Report of the 90-Day Study on Human Exploration of the Moon and Mars," 1989.

National Commission on Space, *Pioneering the Space Frontier* (New York: Bantam Books, 1986).

O'Neill, Gerard K. *The High Frontier* (New York: William Morrow & Co., 1976).

O'Neill, Gerard K. *2081: A Hopeful View of the Human Future* (New York: Simon and Schuster, 1981).

Ride, Sally K., et al, "Leadership and America's Future in Space," NASA, August 1987.

Ryan, Cornelius (ed.). *Across the Space Frontier* (New York: Viking Press, 1952). Contributors included Wernher von Braun, Willy Ley, Fred L. Whipple, and others.

Synthesis Group, "America at the Threshold," White House National Space Council, 1991.

Chapter 3: Muddling Through with a Short-Term View

Baumgartner, Frank & Bryan Jones. *Agendas and Instability in American Politics* (Chicago: University of Chicago Press, 1993).

Burns, James MacGregor. *Leadership* (New York: Harper & Row, 1978).

Carson, Rachel. *Silent Spring* (New York: Houghton Mifflin Co., 1962).

Dean, Cornelia, "Scientific Savvy? In U.S., Not Much," *New York Times*, August 30, 2005.

Esser, Frank & Bernd Spanier, "Picture Perfect News: Sound Bites and Image Bites in American, British, French, and German Elections in a Time Perspective," Institute of Mass Communication and Media Research, University of Zurich, Switzerland, May 2008 (http://www.allacademic.com/one/www/www/index.php?cmd=Download+Document&key=unpublished_manuscript&file_index=1&pop_up=true&no_click_key=true&attachment_style=attachment&PHPSESSID=2a6c8b8ee2ea3f523ac6d74737574807).

Fries, Sylvia Doughty, "Opinion Polls and the U.S. Civil Space Program," paper presented to the American Institute for Aeronautics and Astronautics, April 29, 1992, by the NASA Office of Special Services.

Hall, Richard, "Participation, Abdication, and Representation in Congressional Committees" in L. Dodd & B. Oppenheimer (eds.), *Congress Reconsidered* (Washington: Congressional Quarterly Press, 1993).

Ignatius, David, "Figuring Out Our News Future," *Washington Post*, May 10, 2009, p. A19 (http://www.washingtonpost.com/wp-dyn/content/article/2009/05/08/AR2009050802385.html).

Jones, Alex. *Losing the News: The Future of the News that Feeds Democracy* (New York: Oxford University Press, 2009).

Katz, James E. *Presidential Politics and Science Policy* (New York: Praeger Publishers, 1978).

Kohut, A. & L. Hugick, "20 Years After Apollo 11, Americans Question Space Program's Worth," Gallup Report, No. 286, July 1989.

Lambright, W. Henry, "Managing America to the Moon: A Coalition Analysis" in Pamela E. Mack (ed.), *From Engineering Science to Big Science: The NACA and NASA Collier Trophy Research Project Winners* (Washington: NASA History Office, SP-4219, 1998).

Light, Paul C. *The President's Agenda* (Baltimore: Johns Hopkins University Press, 1983).

Logsdon, John M. *The Decision to Go to the Moon* (Chicago: University of Chicago Press, 1970).

Miller, Jon D. "The Information Needs of the Public Concerning Space Exploration: A Special Report to the National Aeronautics and Space Administration," Chicago Academy of Sciences, 1994.

Rose, Richard. *The Postmodern President* (Chatham, NJ: Chatham House Publishers, 2nd ed., 1991).

U.S. Senate, "Organization of the Congress," Senate Report No. 103-215, Vol. 1, 1993.

Wildavsky, Aaron, "The Two Presidencies" *Transaction*, Vol. 4, 1966, pp. 7-14.

Wildavsky, Aaron. *The New Politics of the Budgetary Process* (Boston: Scott, Foresman and Company, 1988).

Young, A. Thomas, et al, "Leadership, Organization, and Management for National Security Space," Institute for Defense Analyses, July 2008.

Chapter 4: The Bureaucracy: Best Hope for the Future?

Bilstein, Roger E. *Orders of Magnitude: A History of the NACA and NASA, 1915-1990* (Washington: NASA History Series, SP-4406, 1989).

Bush, George W., "U.S. National Space Policy," National Security Presidential Directive (NSPD) 49, August 31, 2006.

Communications Satellite Act of 1962, Public Law No. 87-624, August 31, 1962.

Goodsell, Charles T. *The Case for Bureaucracy* (Washington: Congressional Quarterly Press, fourth edition, 2004).

NASA History Division, "Electronics Research Center," *http://history.nasa.gov/erc.html*.

NASA Institute for Advanced Concepts, "9th Annual and Final Report, 2006-2007" (*http://www.niac.usra.edu/files/library/annual_report/2006annualreport.pdf*).

National Aeronautics and Space Act of 1958, as amended (NASA charter), Public Law No. 85-568 (*http://history.nasa.gov/spaceact-legishistory.pdf*).

U.S. National Research Council, Aeronautics and Space Engineering Board, "Fostering Visions for the Future: A Review of the NASA Institute for Advanced Concepts," Committee to Review the NASA Institute for Advanced Concepts, National Academies Press, 2009 (*http://www.nap.edu/catalog/12702.html*).

Chapter 5: Astropreneurs: The Real Vision, or Just a Dream with Good Special Effects?

Aldridge, Edward C., et al., "Report of the President's Commission on Implementation of United States Space Exploration Policy: A Journey to Inspire, Innovate, and Discover," U.S. Government Printing Office, June 2004 (*http://www.nasa.gov/pdf/60736main_M2M_report_small.pdf*).

Bush, George H.W., National Security Directive 30 / National Space Policy Directive 1, "Fact Sheet: U.S. National Space Policy," White House Office of the Press Secretary, November 2, 1989.

Reagan, Ronald, "Fact Sheet: Presidential Directive on National Space Policy," National Security Decision Directive 293, February 11, 1988.

Fortune 500 companies (*http://money.cnn.com/magazines/fortune/fortune500/2009/full_list/*).

Michaud, Michael, *Reaching for the High Frontier: The American Pro-Space Movement, 1972-84* (New York: Praeger Publishers, 1986).

NASA and the Space Transportation Association, "General Public Space Travel and Tourism," March 25, 1998.

Space Foundation, *The Space Report 2009* (*http://www.thespacereport.org/store/*).

Satellite Industry Association, "State of the Satellite Industry Report," June 2009 (*http://www.sia.org/news_events/2009_State_of_Satellite_Industry_Report.pdf*).

U.N. Agreement Governing the Activities of States on the Moon and Other Celestial Bodies (Moon Treaty), 1979 (*http://www.oosa.unvienna.org/oosa/en/SpaceLaw/moon.html*).

U.N. Treaty on Principles Governing the Activities of States in the Exploration and Use of Outer Space, including the Moon and Other Celestial Bodies (Outer Space Treaty), 1967 (*http://www.oosa.unvienna.org/oosa/en/SpaceLaw/outerspt.html*).

Chapter 6: Be Careful What You Wish For

Associated Press, "Boeing Fined Over Export of Products With Military Uses," *Los Angeles Times*, April 10, 2006.

Center for Strategic and International Studies (CSIS), "Preserving America's Strength in Satellite Technology," CSIS Satellite Commission, April 2002 (*http://csis.org/files/media/csis/pubs/081023_lewis_satellitetech.pdf*).

Center for Strategic and International Studies (CSIS), "The Health of the U.S. Space Industrial Base and the Impact of Export Controls,"

February 2008 (*http://csis.org/files/media/csis/pubs/021908_csis_spaceindustryitar_final.pdf*).

Clinton, William J., Presidential Decision Directive 49, "Fact Sheet: National Space Policy," White House National Science and Technology Council, September 19, 1996.

Diamond, Jared. *Collapse: How Societies Choose to Fail or Succeed* (New York; Viking Press, 2005).

Johnson-Freese, Joan. "Alice in Licenseland: US satellite export controls since 1990," *Space Policy*, Volume 16, Issue 3, July 2000, pp. 195-204.

Logsdon, John M., "The Decision to Develop the Space Shuttle," *Space Policy*, Vol. 2, No. 2, 1986, pp. 103-119.

Moorman, Thomas S., et al., "U.S. Defense Industry Under Siege—An Agenda for Change," Booz Allen & Hamilton, 2000 (*http://www.boozallen.com/media/file/80445.pdf*).

Reagan, Ronald, National Security Study Directive 5-83, "Space Station," April 11, 1983.

Satellite Industry Association, "Study Shows California Satellite Manufacturers Lost Revenue and Jobs to Foreign Competitors in 2000—Export Controls Cited as Major Factor," news release, February 6, 2001.

Zelnio, Ryan. "The effects of export control on the space industry," *The Space Review*, January 16, 2006 (*http://www.thespacereview.com/article/533/1*).

Chapter 7: Earth as an Open System

Coalition for Space Exploration, summaries of public opinion polls (*http://www.spacecoalition.com/Gallup_Polls.cfm*).

Harris Poll #30, "Closing the Budget Deficit: U.S. Adults Strongly Resist Raising Any Taxes Except 'Sin Taxes' Or Cutting Major Programs," April 10, 2007 (*http://www.harrisinteractive.com/harris_poll/printerfriend/index.asp?PID=746*).

Lewis, John & Ruth. *Space Resources: Breaking the Bonds of Earth* (New York: Columbia University Press, 1987).

Marburger, John (science advisor to President G.W. Bush), keynote speech to the American Astronautical Society's 44th Robert H. Goddard Memorial Symposium, Greenbelt, MD, March 15, 2006 (*http://www.spaceref.com/news/viewsr.html?pid=19999*).

Maslow, Abraham H., "A Theory of Human Motivation," *Psychological Review* 50(4), 1943, pp. 370-396 (*http://psychclassics.yorku.ca/Maslow/motivation.htm*).

Tyson, Neil deGrasse, "Space Exploration: The Power of the Vision," in Space Foundation, *America's Vision: The Case for Space Exploration*, March 2006, pp. 2-7 (*http://www.spacefoundation.org/docs/The_Case_For_Space_Exploration.pdf*).

U.S. Census Bureau, International Data Base, updated December 15, 2008 (*http://www.census.gov/ipc/www/idb/worldpop.html*).

Chapter 8: The Century Perspective

Center for Naval Analysis, Military Advisory Board, "Powering America's Defense: Energy and the Risks to National Security," May 2009 (*http://cna.org/nationalsecurity/energy/*).

Defense Science Board, Task Force on DoD Energy Strategy, "More Fight, Less Fuel," U.S. Department of Defense, February 2008 (*http://www.acq.osd.mil/dsb/reports/2008-02-ESTF.pdf*).

Glennon, Robert. *Unquenchable: America's Water Crisis and What To Do About It* (Washington: Island Press, 2009).

Grisham, L.R., "Nuclear Fusion" in Letcher, Trevor M. (ed.) *Future Energy: Improved, Sustainable, and Clean Options for Our Planet* (Oxford, UK: Elsevier Ltd., 2008), pp. 291-302.

Mankins, John, "A Fresh Look at Space Solar Power: New Architectures, Concepts, and Technologies," Advanced Projects Office, National Aeronautics and Space Administration, October 6, 1997.

Milstein, Mati, "Desalination No 'Silver Bullet' in Mideast," *National Geographic News*, May 22, 2008 (*http://news.nationalgeographic.com/news/2008/05/080522-middle-east.html*).

National Security Space Office, "Space-Based Solar Power as an Opportunity for Strategic Security," U.S. Department of Defense, October 10, 2007 (*http://www.acq.osd.mil/nsso/solar/SBSPInterimAssesment0.1.pdf*).

Organization for Economic Cooperation and Development, "Space 2030: Exploring the Future of Space Applications," May 3, 2004; and "Space 2030: Tackling Society's Challenges," May 31, 2005 (*http://www.oecdbookshop.org*).

Schirber, Michael, "Why Desalination Doesn't Work (Yet)," *LiveScience*, June 25, 2007 (*http://www.livescience.com/environment/070625_desalination_membranes.html*).

Science Applications International Corporation (SAIC), "Public Transportation's Contribution to U.S. Greenhouse Gas Reduction," September 2007 (*http://www.apta.com/resources/reportsandpublications/Documents/climate_change.pdf*).

Sierra Club, "The Dirty Truth About Coal: Why Yesterday's Technology Should Not Be Part of Tomorrow's Energy Future," June 2007 (*http://www.sierraclub.org/coal/dirtytruth/coalreport.pdf*).

Sierra Club & Worldwatch Institute, "Smart Choices for Biofuels," January 2009 (*http://www.sierraclub.org/transportation/downloads/biofuels.pdf*).

Sweet, Phoebe, "Desalination gets a serious look," *Las Vegas Sun*, March 21, 2008 (*http://www.lasvegassun.com/news/2008/mar/21/desalination-gets-serious-look/*).

U.S. Department of Energy, Energy Information Administration, "International Energy Outlook 2009," DOE/EIA-0484(2009), May 27, 2009 (*http://www.eia.doe.gov/oiaf/ieo/index.html*).

U.S. Global Climate Change Program, "Global Climate Change Impacts in the United States," Thomas R. Karl, Jerry M. Melillo, & Thomas C. Peterson (eds.), Cambridge University Press, 2009 (*http://www.globalchange.gov/usimpacts*).

U.S. National Research Council, "Desalination: A National Perspective," 2008 (*http://www.nap.edu/catalog.php?record_id=12184*).

Chapter 9: Commitment to the Future

Ley, Willy. *Beyond the Solar System* (New York: Viking Press, 1964).

Lempert, Robert J., Steven W. Popper, & Steven C. Bankes, "Shaping the Next One Hundred Years: New Methods for Quantitative, Long-Term Policy Analysis," RAND Frederick S. Pardee Center for Longer Range Global Policy and the Future Human Condition, 2003 (*http://www.rand.org/pubs/monograph_reports/MR1626/*).

U.S. National Research Council, Space Studies Board, Committee on the Rationale and Goals of the U.S. Civil Space Program, "America's Future in Space: Aligning the Civil Space Program with National Needs," National Academies Press, 2009 (*http://www.nap.edu/catalog/12701.html*).

CPSIA information can be obtained at www.ICGtesting.com
Printed in the USA
LVOW06s0432190813

348463LV00002B/617/P

9 781450 013475